T0296867

Cambridge Elements ≡

Elements of Paleontology
edited by
Colin D. Sumrall
University of Tennessee

EXPANDED SAMPLING ACROSS ONTOGENY IN *DELTASUCHUS MOTHERALI* (NEOSUCHIA, CROCODYLIFORMES)

Revealing Ecomorphological Niche Partitioning and Appalachian Endemism in Cenomanian Crocodyliforms

Stephanie K. Drumheller
University of Tennessee

Thomas L. Adams
Witte Museum

Hannah Maddox
University of Tennessee

Christopher R. Noto
University of Wisconsin-Parkside

CAMBRIDGE
UNIVERSITY PRESS

University Printing House, Cambridge CB2 8BS, United Kingdom

One Liberty Plaza, 20th Floor, New York, NY 10006, USA

477 Williamstown Road, Port Melbourne, VIC 3207, Australia

314–321, 3rd Floor, Plot 3, Splendor Forum, Jasola District Centre, New Delhi – 110025, India

79 Anson Road, #06–04/06, Singapore 079906

Cambridge University Press is part of the University of Cambridge.

It furthers the University's mission by disseminating knowledge in the pursuit of education, learning, and research at the highest international levels of excellence.

www.cambridge.org
Information on this title: www.cambridge.org/9781009005814
DOI: 10.1017/9781009042024

© Stephanie K. Drumheller, Thomas L. Adams, Hannah Maddox, and Christopher R. Noto 2021

When citing this work, please include a reference to the DOI 10.1017/9781009042024

First published 2021

A catalogue record for this publication is available from the British Library.

ISBN 978-1-009-00581-4 Paperback
ISSN 2517-780X (online)
ISSN 2517-7796 (print)

Additional resources for this publication at www.cambridge.org/drumheller

Expanded Sampling Across Ontogeny in *Deltasuchus motherali* (Neosuchia, Crocodyliformes)

Revealing Ecomorphological Niche Partitioning and Appalachian Endemism in Cenomanian Crocodyliforms

Elements of Paleontology

DOI: 10.1017/9781009042024
First published online: April 2021

Stephanie K. Drumheller
University of Tennessee

Thomas L. Adams
Witte Museum

Hannah Maddox
University of Tennessee

Christopher R. Noto
University of Wisconsin-Parkside

Author for correspondence: Stephanie K. Drumheller, sdrumhel@utk.edu

Abstract: New material attributable to *Deltasuchus motherali*, a neosuchian from the Cenomanian of Texas, provides sampling across much of the ontogeny of this species. Detailed descriptions provide information about the paleobiology of this species, particularly with regards to how growth and development affected diet. Overall snout shape became progressively wider and more robust with age, suggesting that dietary shifts from juvenile to adult were not only a matter of size change, but of functional performance as well. These newly described elements provide additional characters upon which to base more robust phylogenetic analyses. The authors provide a revised diagnosis of this species, describing the new material and discussing incidents of apparent ontogenetic variation across the sampled population. The results of the ensuing phylogenetic analyses both situate *Deltasuchus* within an endemic clade of Appalachian crocodyliforms, separate and diagnosable from goniopholidids and pholidosaurs, herein referred to as Paluxysuchidae. This title is also available as Open Access on Cambridge Core

Keywords: Cretaceous, Woodbine Formation, ontogeny, Crocodyliformes, *Deltasuchus*

ISBNs: 9781009005814 (PB), 9781009042024 (OC)
ISSNs: 2517-780X (online), 2517-7796 (print)

Contents

Further Online Appendices, in addition to other supplementary materials including video files, can be accessed at www.cambridge.org/drumheller.

1 Introduction

Starting in the mid-Cretaceous, the spread of the Western Interior Seaway divided North America in half, an event that should lend itself well to explorations of vicariance in terrestrial and freshwater taxa. However, such analyses are stymied by two separate taphonomic biases. First, the North American record is temporally biased, with spikes in diversity being known in the Aptian–Albian and the Campanian–Maastrichtian with few terrestrial sites in between (Jacobs and Winkler, 1998; Weishampel et al., 2004; Zanno and Makovicky, 2013). Secondly, what mid-Cretaceous sites we do have are geographically biased as well, concentrated on the western landmass of Laramidia (Ullmann et al., 2012; Krumenacker et al., 2016; Prieto-Márquez et al., 2016). Appalachia, to the east, has remained something of a mystery, but recent discoveries are starting to reveal aspects of the diversity of this understudied landmass (e.g. Adams et al., 2017; Brownstein, 2018; Adrian et al., 2019; Noto et al., 2019).

The Woodbine Group outcrops across north-central Texas and is situated within both this temporal and geographic gap. Dating to 96 Ma and situated on the western paleocoastline of Appalachia (Powell, 1968; Dodge, 1969; Kennedy and Cobban, 1990; Emerson et al., 1994; Lee, 1997a, 1997b; Jacobs and Winkler, 1998; Gradstein et al., 2004), the Woodbine long has provided tantalizing hints to Appalachian diversity. Unfortunately, fossils from this unit are often fragmentary and isolated, frustrating taxonomic identification beyond broad groupings of Cretaceous organisms (Lee, 1997a; Head, 1998; Jacobs and Winkler, 1998; Adams et al., 2011). The Arlington Archosaur Site (AAS) is an unusual outlier amidst other Woodbine localities. The quality of preservation and the density of recovered fossil material are both unusually high, providing a window into the paleoecosystem of the Cenomanian coastline (e.g. Adams et al., 2017; Brownstein, 2018; Adrian et al., 2019; Noto et al., 2019).

As its name suggests, the site is particularly rich in archosaurian fossils, with at least four crocodyliform (Adams et al., 2017; Noto et al., 2019) and five dinosaur taxa (Main et al., 2014; Noto, 2016) known from the locality. Among these taxa, the most common species recovered from the AAS is the large neosuchian crocodyliform *Deltasuchus motherali* (Adams et al., 2017). Originally described from a single, adult individual, ongoing research and collection of AAS materials have revealed numerous smaller-bodied specimens attributable to this taxon. Furthermore, searches of museum collections at Southern Methodist University (SMU) and the Witte Museum (WM) have resulted in the identification of additional *D. motherali* elements from Bear Creek (SMU Locality 245) near the south entrance to Dallas–Fort Worth

International Airport. Here we describe juvenile to adult *Deltasuchus* elements, emphasizing ontogenetic change seen across the group. We also expand the original phylogenetic analysis of *D. motherali* with new elements preserved in these additional specimens, and increase taxon sampling to further explore a seemingly endemic, Appalachian radiation of neosuchians.

Anatomical Abbreviations – Alveolar and dental positions are named with the first letter of the supporting bone (**p** for premaxilla, **m** for maxilla, **d** for dentary) and the number of the alveolus or tooth counting from mesial to distal along the toothrow; **ang**, angular; **ar**, articular; **ars**, articular sutural surface; **cqc**, cranio-quadrate canal; **d**, dentary; **ects**, ectopterygoid sutural surface; **fae**, foramen aëreum; **f**, frontal; **fio**, foramen intermandibularis oralis; **fs**, frontal sutural surface; **gf**, glenoid fossa; **ics**, intercondylar sulcus; **j**, jugal; **js**, jugal sutural surface; **lac**, lacrimal; **lac no**, lacrimal notch; **lac plp**, lacrimal posterolateral process; **lacs**, lacrimal sutural surface; lacrimal; **lac js**, lacrimal-jugal sutural surface; **lhc**, lateral hemicondyle; **mAME**, M. adductor mandibulae externus; **mhc**, medial hemicondyle; **mx**, maxilla; **mxs**, maxillary sutural surface; **nar**, naris; **nas**, nasal sutural surface; **op**, occlusal pit; **par**, parietal; **pars**, parietal sutural surface; **parops**, paroccipital process sutural surface; **pmx**, premaxilla; **pmx mxs**, premaxillary-maxillary sutural surface; **pob**, postorbital bar; **pos**, postorbital sutural surface; **prf**, prefrontal; **prfp**, prefrontal pillar; **prfs**, prefrontal sutural surface; **q**, quadrate; **qad**, quadrate anterodorsal process; **qat**, adductor tubercle of the quadrate; **qd**, quadrate dorsal process; **qjs**, quadrato-jugal sutural surface; **qpt**, quadrate pterygoid process; **qvf**, ventral fossa of the quadrate; **qs**, quadrate sutural surface; **ra**, retroarticular process; **roe**, external otic recess; **sp**, splenial; **sps**, splenial sutural surface; **sq**, squamosal; **sqj**, spina quadratojugalis; **sqs**, squamosal sutural surface; **sur**, surangular; **surs**, surangular sutural surface; **sym**, mandibular symphysis; **uef**, groove for upper ear valve.

Institution Abbreviation – **DMNH**, Perot Museum of Nature and Science, Dallas, Texas, USA; **SMU**, Southern Methodist University Shuler Museum of Paleontology, Dallas, Texas, USA; **WM**, Witte Museum, San Antonio, Texas, USA.

2 Age and Geologic Setting

The Upper Cretaceous (middle to upper Cenomanian) Woodbine Group of Texas represents a series of fully marine to terrestrial rocks deposited as a southward thinning clastic wedge in the Gulf Coast Basin (Figure 1) (Dodge, 1952, 1969; Oliver, 1971). Recent studies based on subsurface data

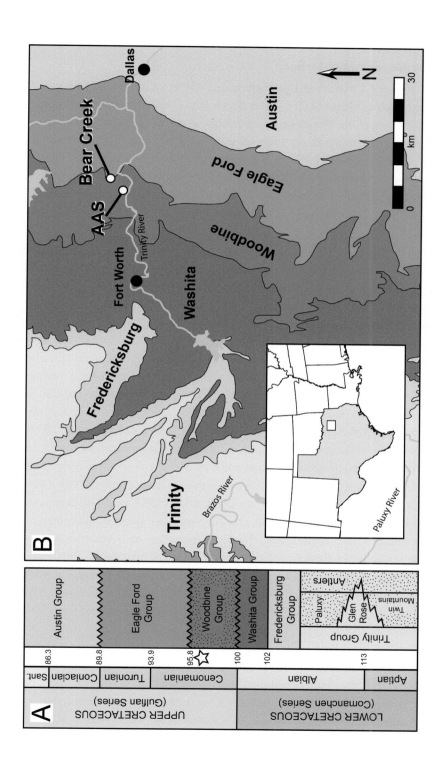

Caption for Figure 1 (cont.)

Figure 1 Location and geologic setting of the AAS and Bear Creek. **A,** stratigraphic column for the Upper Cretaceous of north-central Texas showing the position of the Woodbine Group relative to timescale and adjacent geologic units. Stippled intervals represent terrestrial deposits. Star indicates position of the AAS and Bear Creek sites. Time scale based on Gradstein et al. (2004). **B,** generalized map of geologic units in the Fort Worth Basin, showing enlarged area from white box of inset map of Texas. Modified from Barnes et al. (1972) and Strganac (2015).

classify the Woodbine as a third-order regressive sequence deposited over ~1.5 million years, with source sediments originating from the Ouachita and Arbuckle Mountains of Oklahoma and Arkansas (Ambrose et al., 2009; Adams and Carr, 2010; Blum and Pecha, 2014; Hentz et al., 2014). Its lower boundary is formed by an unconformity with the Grayson Marl (Washita Group) while its upper boundary is formed by an unconformity with the Eagle Ford Group (Dodge, 1969; Oliver, 1971; Johnson, 1974). Presence of the ammonite zonal marker *Conlinoceras tarrentense* in the upper Woodbine establishes a minimum age of early middle Cenomanian (~96 million years) (Kennedy and Cobban, 1990; Emerson et al., 1994; Jacobs and Winkler, 1998; Gradstein et al., 2004), with deposition ending no later than 92 million years (Ambrose et al., 2009).

Woodbine stratigraphy is complex, with differing interpretations arising from surface outcrop and subsurface core and wireline data, themselves derived from widely different locations within the depositional basin (Dodge, 1969; Oliver, 1971; Johnson, 1974; Ambrose et al., 2009; Adams and Carr, 2010; Hentz et al., 2014). In the Dallas–Fort Worth area four units are typically recognized (in ascending order): Rush Creek, Dexter, Lewisville, and Arlington (Dodge, 1969). The lower Rush Creek and Dexter represent marginal to fully marine deposits, while the upper Lewisville and Arlington represent terrigenous fluvio-deltaic environments all influenced by eustatic sea-level changes (Powell, 1968; Oliver, 1971; Johnson, 1974; Ambrose et al., 2009; Adams and Carr, 2010; Hentz et al., 2014).

The Arlington Archosaur Site (AAS) is situated within the upper Woodbine (Lewisville) and consists of a 200 m long, 5 m thick hillside exposure, with 50 m representing the main fossil quarry (Figure 1B). This outcrop preserves a coastal environment that includes a mixture of marine, freshwater, and terrestrial influences (Noto et al., 2012; Adams et al., 2017; Noto et al., 2019). The outcrop is divided into four distinct facies, recording a transition from primarily terrestrial to primarily marine deposition (Main, 2013; Adams et al.,

2017). The majority of fossils are found in the lowermost facies, which most likely represents a freshwater or brackish wetland deposited within a lower delta plain system (Noto et al., 2012; Main, 2013; Adams et al., 2017). The AAS is remarkable for the sheer number and diversity of organisms recovered (at least 37 unique vertebrate taxa to date, as well as invertebrates and plants) and the quality of preservation of the material (Noto, 2015; Adams et al., 2017).

Bear Creek (SMU locality 245, Figure 1B) is positioned in the uppermost Woodbine (Arlington) and represents the transition phase of the Woodbine Formation into the deeper marine shale facies of the Eagle Ford Group (Lee, 1997a). This site has produced numerous fossil remains, including teeth and bone fragments of crocodyliforms and dinosaurs. The outcrop consists of a lower shaly sandstone unit and an upper sandy shale unit interbedded with thin fossiliferous sandstone that contain dark, lignitic, and carbonaceous layers. Thin units consist of very fine- to fine-grained sandstone with ferruginous cement, iron concretions, chert pebbles, and phosphatic nodules. A phosphatic pebble conglomerate surface, marking a transgressive lag deposit, is rich in reworked vertebrate teeth and small fragments including fishes, frogs, turtles, crocodiles, dinosaurs, and a mammal (Lee, 1997a). The uppermost Woodbine preserves a terrigenous coastal depositional system with fluvio-deltaic influences (Powell, 1968; Dodge, 1969; Lee, 1997a, 1997b; Jacobs and Winkler, 1998) and represents a low-stand sequence within an early trans-gressive system tract of the Greenhorn Cycle of Kauffman and Caldwell (1993).

The taphonomy of these sites is also complex, representing a largely time-averaged assemblage formed through a variety of taphonomic modes, including subaerial exposure, aqueous transport, and predation (Noto et al., 2012; Main et al., 2014; Noto, 2015; Adams et al., 2017). Within the fossil-rich layer of the AAS (Facies A sensu Adams et al., 2017), elements are generally well-preserved, if disarticulated. The presence of mixed marine, brackish, fresh-water, and terrestrial taxa suggest a parautochthonous assemblage, sampled from across the paleodeltaic system, but common vertebrate coprolites, as well as the high quality of the fossil preservation and the minimal spread of associated skeletal elements, indicate that transport was low energy and fairly minimal (McNulty and Slaughter, 1968; Russell, 1988; Cumbaa et al., 2010; Adams et al., 2017; Noto et al., 2019).

Most fossils were collected in situ and these elements are a rich, chocolate brown color. The surface quality is high, and the remains are preserved in three dimensions, but a few exhibit minor compression and distortion (e.g. the left maxilla and the right dentary/splenial of DMNH 2013–07–1859). Elements that were collected at or near the surface have been exposed to more extensive weathering and mineral overgrowth, resulting in changes in color, texture, and overall preservational quality. These elements range from a light brown to light gray color, and take on a chalky appearance (e.g. right premaxilla DMNH 2013–07–1636 and left quadrate DMNH 2013–07–0733). A small number exhibit extensive gypsum overgrowth (e.g. DMNH 2014–06–01, a poorly preserved, left angular).

3 Systematic Paleontology

CROCODYLIFORMES Hay, 1930
MESOEUCROCODYLIA Whetstone and Whybrow, 1983
NEOSUCHIA Benton and Clark, 1988
PALUXYSUCHIDAE, clade nov.

Phylogenetic Definition – Branch-based clade comprising all taxa more closely related to _Paluxysuchus newmani_ Adams, 2013, than to either _Goniopholis crassidens_ Owen, 1841 or _Pholidosaurus schaumburgensis_ Meyer, 1841.

Diagnosis – Members of this clade are mid- to large-sized neosuchian crocodylomorphs diagnosable not by autapomorphies, but instead by a unique combination of characters present in other clades (specifically Goniopholididae and Tethysuchia): platyrostral mesorostrine skull (shared with some goniopholidids); maxilla festooned, with well-defined anterior wave, projecting laterally and ventrally (shared with some goniopholidids); posterior ramus of prefrontal is long, reaching the median region of the orbits (shared with goniopholidids and pholidosaurids); lacrimal reaches the anteroventral margin of the orbit; jugal does not exceed the anterior margin of orbit (shared with tethysuchians); the posterior ramus of the jugal beneath the infratemporal fenestra is rod-shaped (shared with pholidosaurids); median process of the frontal extends anterior to the tip of the prefrontal (shared with tethysuchians); postorbital with anterolateral process present (shared with goniopholidids and pholidosaurids); contact between the descending process of the postorbital and the ectopterygoid; no external mandibular fenestra (shared with some goniopholidids and pholidosaurids); surangular extends to the posterior region of the retroarticular process; retroarticular process facing dorsally and paddle shaped (shared with pholidosaurids).

CROCODYLIFORMES Hay, 1930
MESOEUCROCODYLIA Whetstone and Whybrow, 1983
NEOSUCHIA Benton and Clark, 1988
PALUXYSUCHIDAE, clade nov.

DELTASUCHUS Adams et al., 2017
DELTASUCHUS MOTHERALI Adams et al., 2017

Holotype – DMNH 2013–07–0001, partial skull and mandible.

Referred Material – DMNH 2013–07–1859, partial skull and mandible; DMNH 2014–06–01, partial mandible; DMNH 2013–07–0079, right dentary and maxilla; DMNH 2013–07–0297, left premaxilla, right premaxilla; DMNH 2013–07–1888, right dentary; DMNH 2013–07–0239, left dentary; DMNH 2013–07–0218, right dentary; DMNH 2013–07–1984, right dentary; DMNH 2013–07–0240, left dentary; DMNH 2013–07–0322, left dentary; DMNH 2013–07–0228, left dentary; DMNH 2013–07–0312, right dentary; DMNH 2013–07–0802, right dentary; DMNH 2013–07–0219, left maxilla; DMNH 2013–07–1404d, left prefrontal; DMNH 2013–07–0733, left quadrate; DMNH 2013–07–0084, left lacrimal; DMNH 2013–07–1871, frontal; DMNH 2013–07–1992, left and right quadratojugals, left and right quadrates; DMNH 2013–07–1993, left lacrimal; DMNH 2013–07–1994, partial right exoccipital; DMNH 2013–07–1995, right prefrontal; DMNH 2013–07–1997, right quadrate; DMNH 2013–07–1975, right prefrontal, left jugal; DMNH 2013–07–0178, teeth; SMU 76810, articulated right surangular and angular; WM 2019–15 Ga, left premaxilla; WM 2019–15 Gb, tooth.

Revised Diagnosis – A member of Paluxysuchidae differing from other known neosuchians in having the following unique combination of characters and autapomorphies (* = additional characteristics relative to Adams et al., 2017): anterior premaxilla ventrally directed, overbites the dentary (shared with goniopholidids and tethysuchians); postnarial fossa present on the premaxilla; dual pseudocanines on both the dentary and maxilla; frontal excluded from the orbital margin (an autapomorphy for the group)*; anterolaterally facing margin on the dorsal portion of the postorbital; deep fossa on the ventral surface of the quadrate (shared with some pholidosaurids); medial quadrate condyle expands ventrally, separated from lateral condyle by deep intercondylar sulcus; the mandibular symphysis extends posteriorly to the level of the eighth dentary alveolus.

4 Description

4.1 General Description

The AAS is extremely rich in fossil crocodyliforms, with multiple individuals and at least four and perhaps five taxa known from the site (Adams et al., 2017; Noto et al., 2019). Of these, the best represented taxon is *Deltasuchus motherali*. The holotype of *D. motherali* (DMNH 2013–07–0001) includes associated, but disarticulated, craniomandibular elements ascribable to a large, adult neosuchian crocodyliform. The specimen is incomplete, including both premaxillae, maxillae, and nasals, a left postorbital, a left jugal, a right squamosal, both quadrates, a right otoccipital, the basioccipital, both ectopterygoids, and fragments of the pterygoids and dentaries. Based on a reconstructed cranial length of 800 mm, the total body length of the holotype animal is estimated at between 5.6 and 6 m in length. Unlike other large-bodied crocodyliforms known from the mid-Cretaceous of Texas, who exhibit more specialized slender snouts, e.g. *Terminonaris* and *Woodbinesuchus*, *Deltasuchus* has a robust, broadly triangular snout.

Multiple smaller-bodied individuals ascribable to *D. motherali* are also known from the AAS. By far the most complete specimen, other than the holotype (DMNH 2013–07–0001), belongs to an individual that, when articulated, is roughly half the size of the holotype (DMNH 2013–07–1859). Like the holotype (Adams et al., 2017), DMNH 2013–07–1859 includes associated, but disarticulated, cranial and mandibular elements (Figure 2). However, association of these elements to one individual is justified for the following reasons: adjacent elements articulate well along sutural surfaces, all elements were found in close physical proximity to one another within the same bedding plane, and there is no duplication of right and left elements within the skull. At 440 mm in cranial length (measured from the anteriormost tip of the premaxilla, along the midline, to the posteriormost margin of the skull table), this individual would have been between 3.08 and 3.30 m in total body length (sensu Schmidt, 1944; Bellairs, 1969; Adams et al., 2017).

DMNH 2013–07–1859 exhibits the following unique combination of characters that can diagnose it to Paluxysuchidae: enlarged supratemporal fenestrae; paired pseudocanines in the maxilla (m4 and m5), posterior process of the premaxilla overlaps anterodorsal surface of the maxilla anterolaterally, then transition to a butt joint posteromedially; anterior process of the frontal extends anterior to the tip of the prefrontal. It can be assigned to *D. motherali*, as opposed to *Terminonaris* and *Woodbinesuchus* also present in the Woodbine, based on its more robust, widely triangular snout shape, ventrally directed premaxilla, and the extremity of the expansion of the pseudocanines, as well as the associated bulging of the lateral margins of the maxilla which accommodate that enlarged dentition (Adams et al., 2017).

A second individual (DMNH 2014–06–01) is represented by a partial mandible; including left and right articulars, an articulated left angular and surangular and a disarticulated right angular, surangular, and a small portion of posterior dentary (Figure 3). In addition to the elements on each side articulating with one another, these elements were found in close physical proximity and exhibit matching sizes, though the elements from the left side of the jaw exhibit poorer

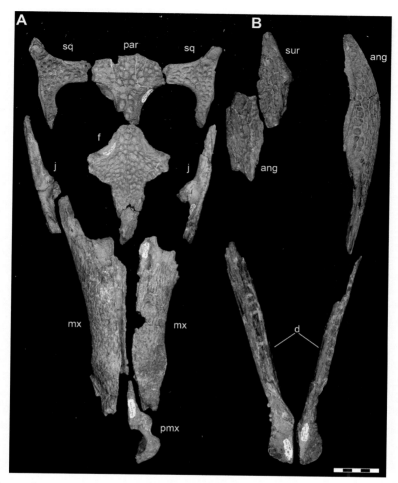

Figure 2 Subadult *D. motherali* (DMNH 2013–07–1859). Cranial elements in **A,** dorsal view and mandible in **B,** dorsal and lateral views. See text for anatomical abbreviations. Scale bar equals 5 cm.

quality surface preservation. This individual would have been of similar size to, and is morphologically indistinguishable from, DMNH 2013–07–1859, but duplication of elements signifies that they do represent two separate animals.

The second set includes a right maxilla and a right anterior fragment of a dentary that were found in direct association with one another (DMNH 2013–07–0079) (Figure 4A–D). This individual exhibits the paired pseudocanines on both maxilla and dentary expected of the clade, and other than being only slightly larger than the other two juveniles previously described, is morphologically similar to these other specimens. Duplication of elements further

Elements of Paleontology

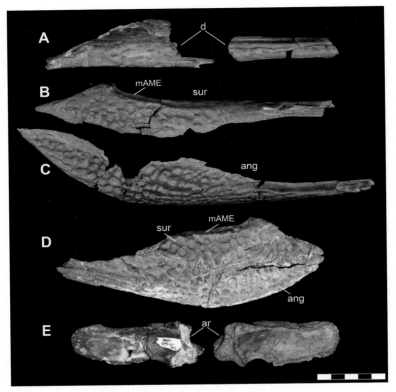

Figure 3 Mandible elements of *D. motherali* (DMNH 2014–06–01). Right dentary in **A**, lateral view; right surangular in **B**, lateral view; right angular in **C**, lateral view; left surangular and angular in **D**, lateral view, dashed line indicating sutural boundary between the angular and surangular; left and right articulars in **E**, dorsal views. See text for anatomical abbreviations. Scale bar equals 5 cm.

differentiates it from DMNH 2013–07–1859. As a comparison of size, the right dentary of DMNH 2013–07–0079 is 34.79 mm wide in ventral view at the level of the d4 alveolus while the right dentary of DMNH 2013–07–1859 is 31.10 mm wide in the same dimension. Scaling the length estimate based on these measurements yields an animal between 3.45 and 3.69 m in total length.

Three larger individuals are represented by isolated elements. DMNH 2013–07–0297 are paired premaxillae that articulate along their sutural margin (Figure 4E). The maximum width of the right premaxillae is 44.70 mm, at the level of p4. When compared to the more complete holotype, DMNH 2013–07–0001, which is 74 mm wide in this dimension, scaling between the two suggest an animal that would have been between 3.38 and 3.62 m in total length. WM

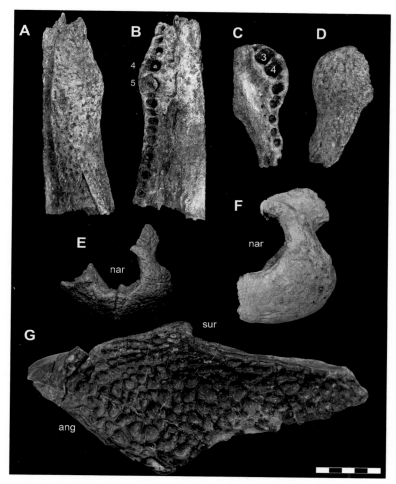

Figure 4 Adult and subadult individuals of *D. motherali*. DMNH 2013–07–0079 right maxillae in **A**, dorsal and **B**, ventral views; right dentary in **C**, dorsal and **D**, ventral views; DMNH 2013–07–0297, premaxillae in **E**, dorsal view; WM 2019–15 Ga, left premaxilla in **F**, dorsal view. SMU 76810, right surangular and angular in **G**, lateral view. See text for anatomical abbreviations. Scale bar equals 5 cm.

2019–15 Ga from Bear Creek is a nearly complete left premaxilla, missing only the narrow posterior process (Figure 4F). The maximum width at the level of p4 is 59.31 mm, indicating a body length of between 4.49 and 4.81 m. The third large individual also comes from Bear Creek. SMU 76810 is an articulated partial right surangular and angular (Figure 4G). It is well-preserved and nearly as large as the surangular of the holotype, DMNH 2013–07–001.

Additionally, at least four left (DMNH 2013–07–0228; DMNH 2013–07–0240 DMNH 2013–07–0322; DMNH 2013–07–0239) (Figure 5A–D) and four right dentaries (DMNH 2013–07–1984; DMNH 2013–07–0218; DMNH 2013–07–0802; DMNH 2013–07–0312) (Figure 5E–H) are known from the site. These dentaries range from highly fragmentary to nearly complete, and all are distinguishable from one another based on size. All exhibit the enlarged d3 and d4 dentition or associated alveoli and the spatulate symphyseal region as seen in this clade. Similar measurements of dentary width at d4

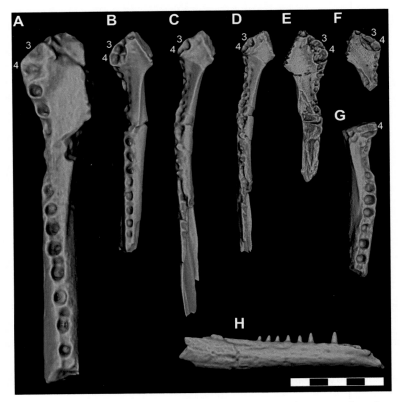

Figure 5 Orthographic image of 3D digital models demonstrating size comparison of dentaries for juvenile and subadult individuals of *D. motherali*. DMNH 2013–07–0239, left dentary in **A**, dorsal view; DMNH 2013–07–0322, left dentary in **B**, dorsal view; DMNH 2013–07–0240, left dentary in **C**, dorsal view; DMNH 2013–07–0228, left dentary in **D**, dorsal view; DMNH 2013–07–1984, right dentary in **E**, dorsal view; DMNH 2013–07–0802, right dentary in **F**, dorsal view; DMNH 2013–07–0218, right dentary in **G**, dorsal view; DMNH 2013–07–0312, right dentary in **H**, lateral view. Scale bar equals 5 cm.

were made to all specimens that included this undamaged portion of bone, yielding the following additional body size estimates: DMNH 2013–07–0228 was 1.38–1.49 m long (based on 13.93 dentary width); DMNH 2013–07–0240 was 1.46–1.56 m long (based on 14.73 mm dentary width); DMNH 2013–07–0802 was 1.65–1.77 m long (based on 16.69 mm dentary width); DMNH 2013–07–0322 was 1.94–2.08 m long (based on 19.57 dentary width); DMNH 2013–07–1984 was 2.03–2.17 m long (based on 20.48 dentary width); DMNH 2013–07–0239 was 2.52–2.70 m (based on 33.07 mm dentary width).

Taken all together, this yields a minimum estimate of *Deltasuchus* individuals currently known from the upper Woodbine of 15. These animals ranged from just under 1.5 m long to roughly 6 m in total length, providing a cross section of individuals across much of the ontogenetic history of the group (Figure 6). In fossil crocodyliforms, defining maturity often relies on patterns of sutural closure, especially in vertebrae (Brochu, 1996; Ikejiri, 2012). Sutural closure in other elements across the crocodyliform body, especially within the skull, can only provide broad brushstroke patterns of maturity during earliest development, with significant variation observed even within extant taxa (Bailleul et al., 2016). While crocodyliform vertebrae are known from the AAS and Bear Creek, as are long bones that could be used for histological analysis, these postcranial elements are not easily associated with any one of the several taxa present at these sites. That leaves our discussion of ontogeny limited to craniomandibular elements. Given the sizes and patterns of sutural closures present in all the specimens described herein, we are confident that we

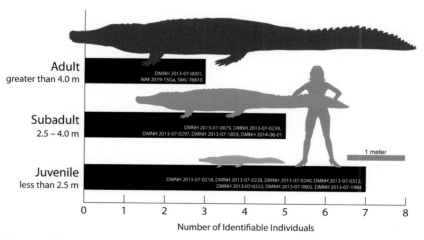

Figure 6 Size comparison of juvenile (less than 2.5 m), subadult (2.5 to 4.0 m), and adult (greater than 4.0 m) individuals of *D. motherali*. Scale bar equals 1 m.

do not have any identifiable neonate remains of *Deltasuchus*. For the sake of comparison and discussion, we will therefore group the individuals estimated to be less than 2.5 m in length as juveniles, the 2.5 to 4 m-individuals as subadults, and anything above 4 m long as an adult animal.

Except where indicated, the element descriptions below refer to the most complete subadult individual, DMNH 2013–07–1859. Several of these elements were not preserved or were fragmentary in the holotype specimen. Other specimens are included to capture morphologies that are not represented in this animal as well as aspects of ontogenetic change observable in each element.

4.2 Premaxilla

DMNH 2013–07–1859 (Video 1) preserves a complete left premaxilla, while DMNH 2013–07–0297 (Video 2) represents a partial left and right premaxillae preserved in three pieces and WM 2019–15 Ga is a large, left premaxilla missing only the posterior process (Figure 7). The anterior terminus of the premaxilla is transversely broad and strongly deflected ventrally, reminiscent of *Sarcosuchus* (Sereno et al., 2001), *Kaprosuchus saharicus* (Sereno and Larsson, 2009), *Elosuchus* and most members of Goniopholididae (Meunier and Larsson, 2017; Jouve et al., in press), so that the premaxilla occludes anterior to the anterior dentary (Figure 8). This beak-like overbite previously has been interpreted as an adaptation for stabilizing and aligning jaw closure in taxa with elongated snouts with powerful bite forces (e.g. *Oceanosuchus* in Hua

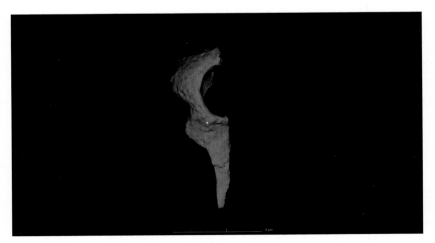

Video 1 Orthographic digital model of the left premaxilla from a subadult *D. motherali*, DMNH 2013–07–1859. Video available at www.cambridge.org/drumheller.

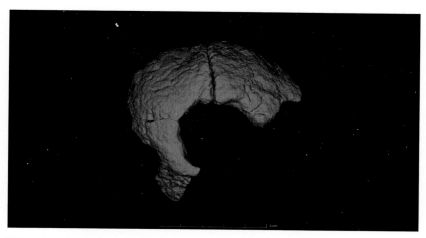

Video 2 Orthographic digital model of the articulated premaxillae from an adult *D. motherali*, DMNH 2013–07–0297. Video available at www.cambridge.org/ drumheller.

et al., 2007). While adult *Deltasuchus* did not have a particularly narrow snout as compared to *Terminonaris* and *Woodbinesuchus* (see Section 6, Discussion), younger members of this taxon, as well as members of its sister taxon, *Paluxysuchus* (Adams, 2013), did have comparatively narrow jaws. The presence of this feature is therefore likely due to a combination of factors: (1) inheritance of a plesiomorphic condition, as many taxa in this region of the tree, especially the goniopholidids *Anteophthalmosuchus epikrator*, *Amphicotylus stovalli* and *Goniopholis kiplingi* and the pholidosaurs *Chalawan*, *Sarcosuchus*, and *Terminonaris*, share this trait (Sereno et al., 2001; Wu et al., 2001; Andrade et al., 2011; Martin et al., 2014; Young et al., 2016; Ristevski et al., 2018); and (2) ontogenetic inertia (sensu Gignac and O'Brien, 2016), given that the stabilizing aspects of this trait would have proven more advantageous to the more slender-snouted juveniles than the broader-faced adults.

In lateral view, the premaxilla is triangular-shaped, narrowing and thinning posteriorly to become a flattened, posterior process (Figure 7C, D). Just anterior to the premaxilla-maxillary contact is a large, lateral notch for the reception of the enlarged d4 and d5 pseudocanines. When articulated, the premaxillae completely enclose the dorsally oriented external naris. There are small, posteriorly directed processes on the anteromedial margin of the naris as seen in the holotype (DMNH 2013–07–0001). The dorsal margin of the narial rim is elevated sloping posteriorly to the level of the maxilla. In dorsal view, the posterior process of the premaxilla narrows to a point (Figure 7A, B). The premaxillae would articulate at the midline along an edge-to-edge suture to

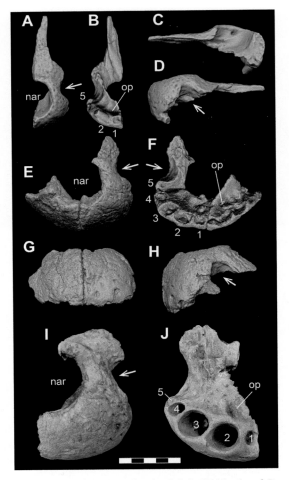

Figure 7 Premaxillae of adult and subadult individuals of *D. motherali*.
Orthographic image of 3D digital models of DMNH 2013–07–1859 in **A**,
dorsal, **B**, ventral, **C**, medial, and **D**, lateral views; orthographic image of 3D
digital models of DMNH 2013–07–0297 in **E**, dorsal, **F**, ventral, **G**, anterior,
and **H**, lateral views; WM 2019–15 Ga in **I**, dorsal and **J**, ventral views. Lateral
notch for d4 and d5 occlusal indicated by open-headed arrows. See text for
anatomical abbreviations. Scale bar equals 5 cm.

wedge between the maxillae and extends posteriorly to contact the nasal at the
level of m5. The premaxillae exclude the maxillae and nasals from contacting
the posterior margin of the naris. In palatal view, five premaxillary alveoli are
arranged along the anterior margin, with p2 and p3 being larger than p1 and p4.
P5 is not preserved in the holotype DMNH 2013–07–0001 but is present in

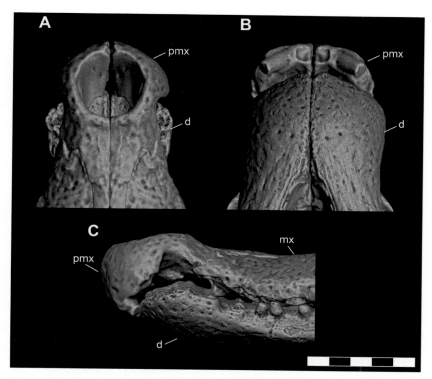

Figure 8 Orthographic image of 3D digital model reconstruction of the subadult *D. motherali* (DMNH 2013–07–1859) demonstrating premaxilla-dentary occlusion and the ventral reflection of the premaxilla in **A**, dorsal, **B**, ventral, and **C**, lateral views. See text for anatomical abbreviations. Scale bar equals 5 cm.

DMNH 2013–07–0297 and WM 2019–15 Ga as an extremely reduced alveolus directly posterior to p4. P1 through p3 extend ventrally to the same level, while p4 and p5 occur higher in a step wise fashion so that p5 is in line with the palate. The palatal surface is deeply concave with pits for the occlusion of the anterior dentary teeth posterior to the ventrally projected anterior premaxillary teeth (Figure 7B, F, J).

4.3 Maxilla

DMNH 2013–07–1859 preserves both left and right maxillae (Figure 9A–D; Video 3). The right is nearly complete while the left is missing portions of the palate and posterior maxillary process. Two additional partial maxillae (DMNH 2013–07–0079 and DMNH 2013–07–0219) have also been recovered (Figure 9E–I). The dorsal surface is ornamented with shallow pits and grooves but is not as rugose as seen in the adult holotype DMNH

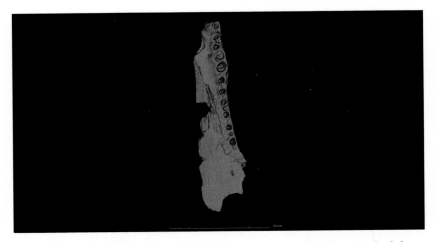

Video 3 Orthographic digital model of the left maxilla from a subadult *D. motherali*, DMNH 2013–07–1859. Pathological puncture visible on the dorsal surface of the rostrum. Video available at www.cambridge.org/drumheller.

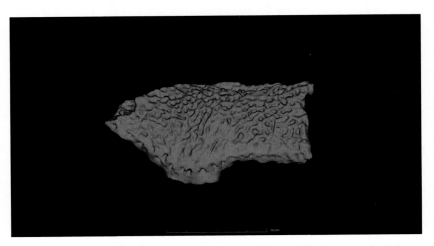

Video 4 Orthographic digital model of the left maxilla from a subadult *D. motherali*, DMNH 2013–07–1975. Video available at www.cambridge.org/drumheller.

2013–07–0001 (Video 4). In dorsal view, the lateral borders of the maxillae are sinusoidal anteriorly and become straighter toward the posterior process. There is a distinct lateral bulge at the level of m4 and m5. Anterior to this bulge, the maxilla tapers toward the premaxilla, resulting in a narrow rostral constriction at the premaxillomaxillary juncture. In lateral

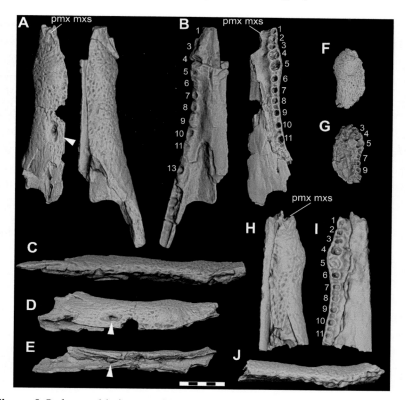

Figure 9 Orthographic image of 3D digital models of maxillae of juvenile and subadult individuals of *D. motherali*. DMNH 2013–07–1859 left and right maxillae in **A**, dorsal, **B**, ventral views; DMNH 2013–07–1859 right maxilla in **C**, lateral view; DMNH 2013–07–1859 left maxilla in **D**, obtuse, **E**, medial views; DMNH 2013–07–0219 in **F**, dorsal, **G**, ventral views; DMNH 2013–07–0079 in **H**, dorsal, **I**, ventral, **J**, lateral views. White arrows indicate a partially healed, bite mark. See text for anatomical abbreviations. Scale bar equals 5 cm.

view, the narrow rostral constriction is upturned dorsally at approximately 17° (Figure 9J). Above the alveolar margin, neurovascular foramina are evenly spaced linearly along the length of the maxilla. The maxilla has an edge-to-edge contact with the nasal along its straight dorsomedial margin. The posteromedial margin of the maxilla is overlapped dorsolaterally by the anterolateral portion of the lacrimal and the anterior process of the jugal. As seen with the holotype (DMNH 2013–07–0001), the posterior maxillary process passes lateral to the orbits and ventromedially to the jugal and contacts the anterior process of the ectopterygoid just anterior to the post-orbital bar.

The secondary palate is well-preserved in DMNH 2013–07–1859 and DMNH 2013–07–0079 with some minor crushing of the palatal shelves medial to the alveolar margin (Figure 9B). Posteriorly, the maxillary palatal process forms the anterior border of the suborbital fenestra and the posterior maxillary process forms most of the anterolateral border. There are 20 maxillary alveoli in the right maxilla, with 11 preserved in the left of DMNH 2013–07–1859. The alveoli increase in size from the first alveolus to m4 and m5, which are the largest. The diameter of alveoli then decreases in size starting with m6 and remain relatively constant in size thereafter. All maxillary alveoli are separated by septa. In between the alveoli are small pits for the occlusion of the dentary teeth.

4.4 Lacrimal

A complete, left lacrimal (DMNH 2013–07–0084) (Figure 10A, B) and a second, incomplete left lacrimal (DMNH 2013–07–1993) were found in isolation, broadly associated with other *D. motherali* material. The element is anteriorly elongated, narrowing to insert between the maxilla and nasal. It meets the prefrontal posteromedially along an edge-to-edge suture, resulting in the lacrimal extending further anteriorly than the prefrontal. The anteromedial border of the lacrimal broadly contacts the nasal and the anterolateral border overlaps the maxilla. The posterior margin of the lacrimal forms most of the anterior and anterolateral border of the orbit. On the dorsal surface, an elevated rim is present on the anteromedial margin of the orbit. Lateral to this rim, a smooth, triangular depression, the lacrimal notch, separates the orbital ridge from the posterolateral process of the lacrimal. The narrow posterolateral process extends to fit into a notch on the anterodorsal tip of the jugal.

4.5 Prefrontal

A complete left prefrontal (DMNH 2013–07–1404) and two partial right prefrontals (DMNH 2013–07–1975 and DMNH 2013–07–1995) were also found isolated (Figure 10C–E). A second partial left prefrontal is fused with a partial frontal (DMNH 2013–07–1871) (Figure 11E–H). DMNH 2013–07–1404 does articulate with the frontal of DMNH 2013–07–1859 (Figure 11A–D). In dorsal view, the prefrontal extends anteriorly to a wedge-shaped process and does not contact the maxilla. The posteromedial margin of the prefrontal is deflected for an oblique articulation in the inset groove in the anteromedial corner of the frontal. As a result, the frontal is completely excluded from the medial margin of the orbit. A curved,

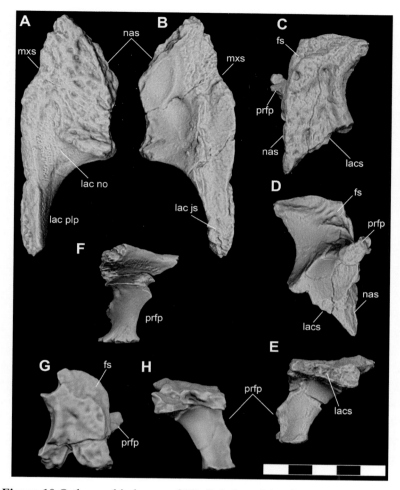

Figure 10 Orthographic image of 3D digital models of cranial elements of subadult individuals of *D. motherali*. Left lacrimal DMNH 2013–07–0084 in **A**, dorsal and **B**, ventral views; left prefrontal DMNH 2013–07–1404 in **C**, dorsal, **D**, ventral and **E**, anterior views; right prefrontal DMNH 2013–07–1975 in **F**, posterior view. Right prefrontal DMNH 2013–07–1995 in **G**, dorsal and **H**, anterior views. See text for anatomical abbreviations. Scale bar equals 5 cm.

elevated rim extends along the posterolateral margin of the prefrontal and becomes confluent with the elevated rim on the lacrimal, forming the supraorbital rim. The medial descending prefrontal pillar is mediolaterally expanded and flat with a small triangular medial process. In ventral view, the base of the pillar is ellipsoid.

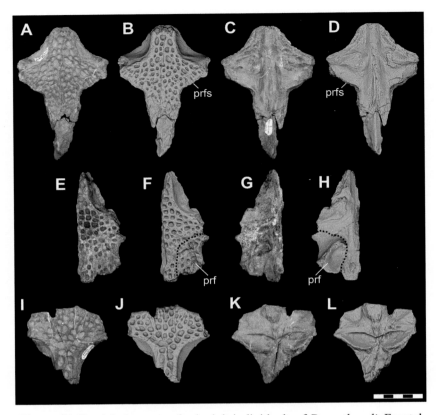

Figure 11 Cranial elements of subadult individuals of *D. motherali*. Frontal DMNH 2013–07–1859 in **A**, dorsal, orthographic image of 3D digital model in **B**, dorsal, **C**, ventral, and orthographic image of 3D digital models of **D**, ventral views; frontal and prefrontal DMNH 2013–07–1871 in **E**, dorsal, orthographic image of 3D digital model in **F**, dorsal, **G**, ventral, and orthographic image of 3D digital models of **H**, ventral views; parietal DMNH 2013–07–1859 in **I**, dorsal, orthographic image of 3D digital model in **J**, dorsal, **K**, ventral, and orthographic image of 3D digital model in **L**, ventral views. See text for anatomical abbreviations. Scale bar equals 5 cm.

4.6 Frontal

The frontals of DMNH 2013–07–1859 are fused. It is a relatively flat and cross-shaped element (Figure 11A–D). The dorsal surface of the frontal is densely ornamented with rounded pits. There is a very slight sagittal ridge extending anteroposteriorly along the length of the dorsal surface. It does not appear to extend to the posterior edge of the preserved frontal. Anterior to the orbits, the median process tapers between the prefrontals to form a wedge between the nasals. This anterior process exceeds the anterior tip of the prefrontal but not the lacrimal. Posterior to the orbits, the frontal expands laterally to meet the postorbital and forms the anteromedial margin of the supratemporal fenestra. The posterolateral margin slopes ventrally to form the supratemporal fossae and excludes the parietal from contact with the postorbital. The posterior process of the frontal contacts the parietal midway between the supratemporal fenestrae in a transverse suture. A descending projection medial to the cristae cranii frontales divides the ventral surface of the anterior median process into two longitudinal grooves. At the level of the postorbitals, the two grooves merge into the single path for the olfactory tract.

A second left frontal, DMNH 2013–07–1871, was also recovered (Figure 11E–H). It is larger than DMNH 2013–07–1858 and is in the range of subadult to adult. Unlike DMNH 2013–07–1858, it is disarticulated from the right frontal and its medial articular margin shows no signs of having been fused at the midline. However, the left prefrontal is completely fused with the frontal with no visible sutural line. This provides strong support to Bailleul et al.'s (2016) argument that cranial sutural closure is not a valid means for assessing relative maturity in extant and extinct archosaurs.

4.7 Parietal

The parietal is an unpaired and nearly flat, T-shaped element (Figure 11I–L). Its dorsal surface is sculpted with rounded pits. There is no indication of a sagittal ridge as seen in the frontal. It contacts the frontal anteriorly to form the dorsomedial border of the supratemporal openings. The parietal does not contact the postorbital. The posteromedial margin of the supratemporal opening slopes ventrolaterally to contribute to the supratemporal fossa. The parietal meets the quadrate ventrolaterally. Posteriorly, the parietal expands to meet the squamosal laterally at a parasagittally oriented suture. The transverse posterior margin of the parietal overhangs the occiput and excludes the supraoccipital from the dorsal surface of the cranial table. Lateral to the postparietal the ventral margin of the parietal forms the dorsal surface of the post-temporal fenestra and excludes the squamosal participation in the openings. On the ventral surface, just posterior to the supratemporal fenestra, are two mediolaterally oriented impressions for the middle ear cavity.

4.8 Jugal

The left and right jugals for DMNH 2013–07–1859 are nearly complete (Video 5), while DMNH 2013–07–1975 is a partial left jugal (Figure 12A–E). The jugal of DMNH 2013–07–1859 is remarkably similar to that of *P. newmani*

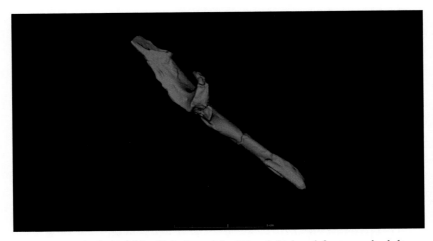

Video 5 Orthographic digital model of the right jugal from a subadult *D. motherali*, DMNH 2013–07–1975. Video available at www.cambridge.org/drumheller.

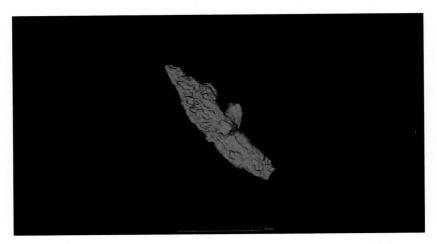

Video 6 Orthographic digital model of the left jugal from an adult *D. motherali*, DMNH 2013–07–0001. Video available at www.cambridge.org/drumheller.

(SMU 76601) (Figure 12F, G). The jugal is an anteroposteriorly elongate, triradiate element. The lateral surface is ornamented with rounded pits and grooves, as in the holotype (Figure 24B; Video 6). There are a series of neurovascular foramina evenly spaced linearly along the length of the lateral margin of the jugal. The anterior and posterior rami are similar in dorsoventral depth. The anterior ramus forms the lateroventral border of the orbit. In cross-

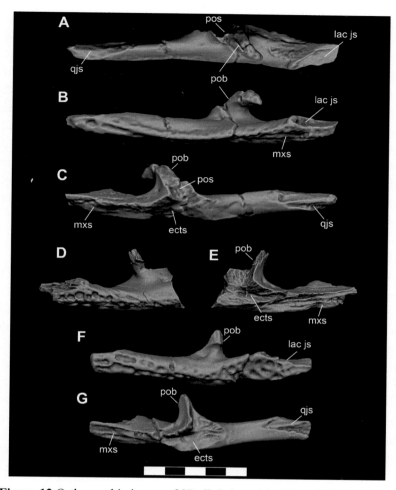

Figure 12 Orthographic image of 3D digital models of cranial elements of subadult individuals of *D. motherali*. Right jugal DMNH 2013–07–1859 in **A**, dorsal, **B**, lateral, and **C**, ventral views; left jugal DMNH 2013–07–1975 in **D**, lateral and **E**, medial views. *P. newmani* right jugal SMU 76601 in **F**, lateral and **G**, medial views. See text for anatomical abbreviations. Scale bar equals 5 cm.

sectional view, the anterior ramus is triangular-shaped and wedges between the posterolateral process of the lacrimal dorsomedially and the maxilla ventrally. The anterodorsal tip has a V-shaped notch for articulation with the posterior process of the lacrimal. The ventral sutural surface with the maxilla is nearly flat and rugose. Ventral to the ascending process of the postorbital bar, the ventro-medial surface is rugose for articulation with the ectopterygoid. The narrow posterior ramus of the jugal is straight and rod-shaped, forming the lateroventral border of the infratemporal fenestra. A longitudinal ridge runs along the lateral surface of the posterior process below the infratemporal fenestra. The posterior end mediolaterally contacts the quadratojugal along an anteroposterior groove. The ascending process of the jugal, which forms the lower portion of the postorbital bar, is inset more medially than dorsally midway along the medial margin of the jugal. The anterior tip of the process is curled anterolaterally and articulates with the descending postorbital process to form a dorsoventrally oriented boss that extends anterolaterally from the postorbital bar. The posteromedial surface of the ascending process is inset and rugose for articulation with ascending process of the ectopterygoid. There are two anteroposteriorly aligned foramina, one at the base of the ascending process and the other directly posterior to the process, along the dorsal surface.

4.9 Squamosal

DMNH 2013–07–1859 preserves both left and right squamosals (Figure 13). The dorsal surface of the squamosal is densely sculpted by rounded pits. The anteromedial edge forms the posterolateral margin of the supratemporal fenestra. The lateral margin of the squamosal narrows anteriorly to underlie the dorsal plate of the postorbital. It is uncertain if it contacts the postorbital bar. The posterior dorsal lamina overhangs the lateral margin of the squamosal and forms the dorsal surface of the external otic recess. The dorsolateral sulcus for the earflap is shallow and extends to the anterolateral margin of the squamosal. The squamosal tapers posterolaterally into a prong like process along the dorsal edge of the paroccipital process and extends beyond the level of the posterior margin of the parietal. The posterior margin is directed medially to contact the parietal and overhang the occiput. Ventrolaterally, it descends to contact the posterodorsal surface of the quadrate posterior to the otic recess and lateral to the paroccipital process. A groove extends along the dorsomedial margin of the paroccipital process, while a groove on the medioventral surface of the paroccipital process forms the roof and lateral wall of the cranioquadrate canal.

4.10 Quadratojugal

DMNH 2013–07–1992 includes nearly complete right and partial left quadratojugals (Figure 14). They are missing anteriormost projection. As a result, it is not possible to determine how far anterior the quadratojugal extends. The spina quadratojugal is prominent and rounded (Figure 4A, B). The dorsal surface is divided into two distinct regions, a smooth, unornamented medial surface and an ornamented lateral region. Medially, the posterolateral margin of the infratemporal opening is smooth and slightly curved anterolaterally. The

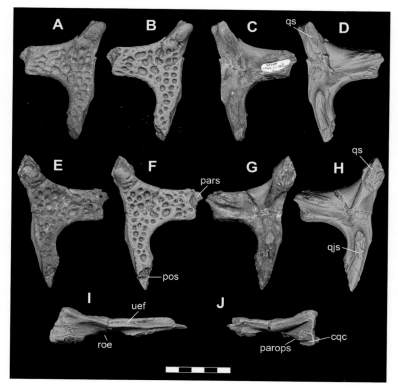

Figure 13 Squamosals of subadult individual of *D. motherali*, DMNH 2013–07–1859. Left in **A**, dorsal; orthographic image of 3D digital model in **B**, dorsal, **C**, ventral; orthographic image of 3D digital model in **D**, ventral views. Right in **E**, dorsal; orthographic image of 3D digital model in **F**, dorsal, **G**, ventral; orthographic image of 3D digital model in **H**, ventral, **I**, lateral, and **J**, posterior views. See text for anatomical abbreviations. Scale bar equals 5 cm.

lobe-shaped posterior region of the quadratojugal is sculpted with large, rounded pits. It runs parallel to the quadrate to reach the posterolateral corner of the lateral hemicondyle of the quadrate to take part in the craniomandibular joint. The ventral surface is smooth and slightly concave with a rugose lateral margin for the overlapping contact with the quadrate.

4.11 Quadrate

DMNH 2013–07–1859 does not preserve either quadrate. However, DMNH 2013–07–1992 includes nearly complete partial left and partial right quadrates

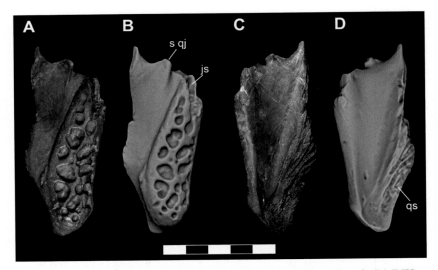

Figure 14 Quadratojugal of subadult individual of *D. motherali*, DMNH 2013–07–1992. Right in **A**, dorsal; orthographic image of 3D digital model in **B**, dorsal, **C**, ventral; orthographic image of 3D digital models in **D**, ventral views. See text for anatomical abbreviations. Scale bar equals 5 cm.

that fits well in the glenoid fossae of the articulars of DMNH 2014–06–01 (Figure 15A–F; Video 7).

An additional, slightly larger, partial left quadrate (DMNH 2013–07–0733) is also known. The anterior half, including the anterodorsal and pterygoid processes, is missing (Figure 15G–I; Video 8).

In all three elements, the quadratojugal articular surface runs lateral to the quadrate to contact the posterolateral corner of the lateral hemicondyle. Medial to that, the dorsal surface is heavily rugose for articulation with the ventral surface of the squamosal and exoccipital. The expanded mandibular condyle is identical to that of the holotype (DMNH 2013–07–0001) (Video 9) with subdivided lateral and medial hemicondyles, separated by a deep intercondylar sulcus. The narrower medial hemicondyle angles medioventrally as compared to the horizontally aligned lateral hemicondyle.

The ventral surface of the quadrate body is dominated by prominent concave crests extending anteriorly from posteromedial margin (Figure 16). In the holotype (DMNH 2013–07–0001), these crests are identified as crest A' and B (sensu Iordansky, 1973) (Figure 16A) and are associated with a deeply recessed fossa. However, this recess is not relatively as deep as that of the holotype, DMNH 2013–07–0001. Like DMNH 2014–06–01,

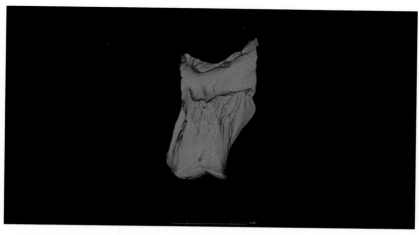

Video 7 Orthographic digital model of the left quadrate from a subadult *D. motherali*, DMNH 2013–07–1992. Video available at www.cambridge.org/drumheller.

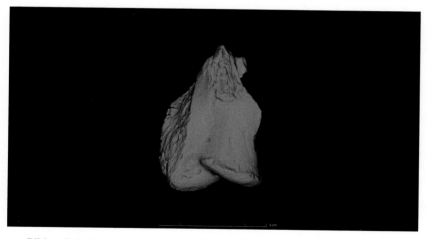

Video 8 Orthographic digital model of the left quadrate from an adult *D. motherali*, DMNH 2013–07–0733. Video available at www.cambridge.org/drumheller.

DMNH 2013–07–0733 and DMNH 2013–07–1992 are most likely subadult individuals half the size of the holotype, suggesting that this deep fossa becomes accentuated during ontogenetic maturity and increased size.

Another partial right quadrate (DMNH 2013–07–1997) is also known (Figure 15J–L). This specimen is roughly half the size of DMNH 2013–07–1992, described above. While this specimen does have a prominent B crest (sensu Adams et al.,

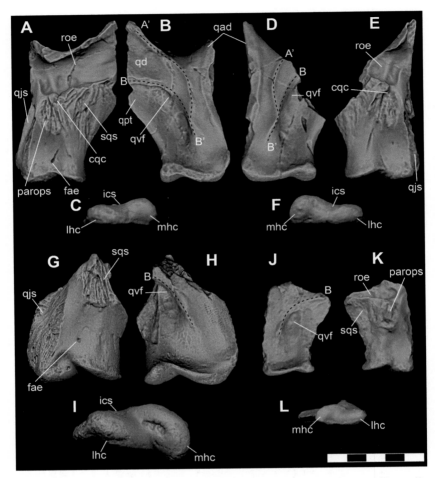

Figure 15 Orthographic images of 3D digital models of quadrates of juvenile and subadult individuals of *D. motherali*. Left quadrate DMNH 2013–07–1992 in **A**, dorsal, **B**, ventral, and **C**, posterior views; right quadrate DMNH 2013–07–1992 in **D**, dorsal, **E**, ventral, and **F**, posterior views; left quadrate DMNH 2013–07–0733 in **G**, dorsal, **H**, ventral and **I**, posterior views; right quadrate DMNH 2013–07–1997 in **J**, dorsal, **K**, ventral, and **L**, posterior views. See text for anatomical abbreviations. Scale bar equals 5 cm.

2017), it lacks the other crests that are present in subadult to adult members of *Deltasuchus*. This raises the possibility that it is attributable to one of the other crocodyliform taxa known from the AAS, for whom quadrates are currently unknown. However, given the more modest crests observed in the subadult specimens described above, it is likely that the prominence of these characters is strongly

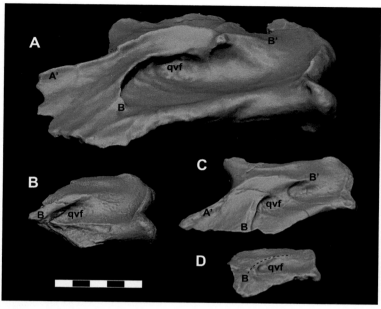

Figure 16 Orthographic images of 3D digital models of juvenile, subadult, and adult individuals of *D. motherali* demonstrating ontogenetic variability in the ventral fossa of the quadrate. Adult right quadrate DMNH 2013–07–0001 (reflected) **A**, subadult left quadrate DMNH 2013–07–0733 **B**, subadult left quadrate DMNH 2013–07–1992 **C**, and juvenile right quadrate DMNH 2013–07–1997 (reflected) **D**, in medial views. See text for anatomical abbreviations. Scale bar equals 5 cm.

affected by ontogeny, with the crests increasing in size to accommodate the larger muscles associated with jaw closure in more mature individuals. Therefore, we tentatively associate this element with a juvenile *Deltasuchus*.

4.12 Dentary

The symphyseal region of the mandibular rostrum is spatulate, with a transversely broad and straight anterior margin (Figure 17). Anteriorly, the mandibular symphysis is low with a shallow, concave dorsal surface. In dorsal aspect, the symphyseal portion of the rostrum extends caudally to a point level with the eighth dentary alveolus. Alveoli d1–d4 are procumbent and transversely aligned, with the fourth alveolus lateral and posterior to the first. Alveoli d3 and d4 are nearly confluent and the largest alveoli of the dentary, resulting in dual pseudo-canines. Small neurovascular foramina occur on the dorsal surface medial to the alveoli. The alveoli are all separated from each other by bony septae.

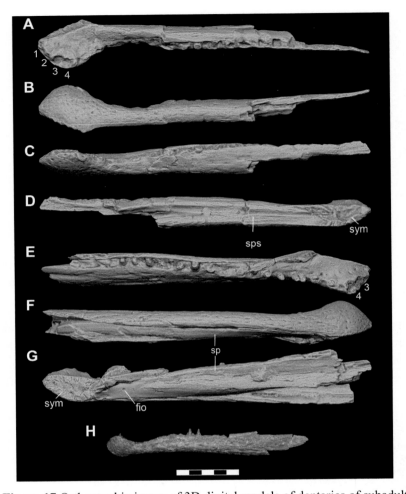

Figure 17 Orthographic image of 3D digital models of dentaries of subadult
individuals of *D. motherali*, DMNH 2013–07–1859. Left dentary in **A**, dorsal,
B, ventral, **C**, lateral, and **D**, medial views; right dentary and splenial in **E**,
dorsal, **F**, ventral, and **G**, medial views. Left dentary of DMNH 2013–07–0240
in **H**, lateral view. See text for anatomical abbreviations. Scale bar equals 5 cm.

The mandibular ramus is long and straight with a convex lateral surface.
The angle at which the ramus diverges from the symphysis varies slightly
with ontogeny, with the smallest specimens (DMNH 2013–07–0228; DMNH
2013–07–1984) (Figure 5G; Video 10, respectively) diverging at an angle of
about 20° and the right dentary of DMNH 2013–07–1859 (the most complete
subadult) diverging at roughly 30° (Figure 2B; Video 11). Without more
complete dentaries in larger individuals, this trend is difficult to map into

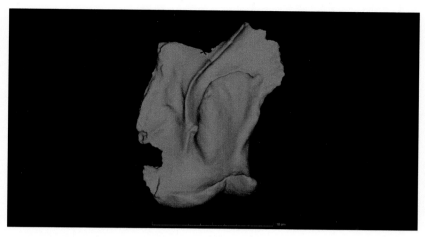

Video 9 Orthographic digital model of the right quadrate from an adult
D. motherali, DMNH 2013–07–0001. Video available at www.cambridge.org/
drumheller.

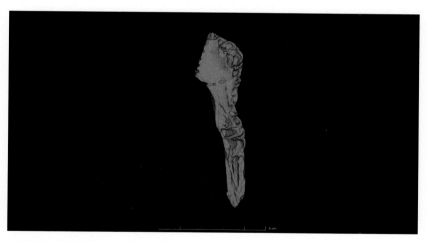

Video 10 Orthographic digital model of the right dentary from a juvenile
D. motherali, DMNH 2013–07–1984. Video available at www.cambridge.org/
drumheller.

adults. However, the reconstructed margins of the rostrum of the holotype
(DMNH 2013–07–0001), which is the largest specimen currently known,
diverge at roughly similar angles to the subadult specimens (Video 12).
During growth, this would be reflected as a widening of the snout as the
animal grew.

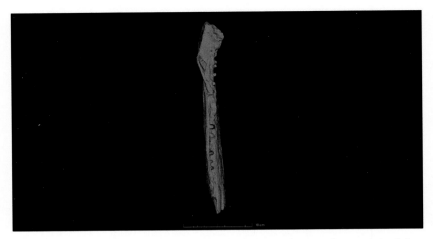

Video 11 Orthographic digital model of the right dentary from a subadult *D. motherali*, DMNH 2013–07–1859. Video available at www.cambridge.org/drumheller.

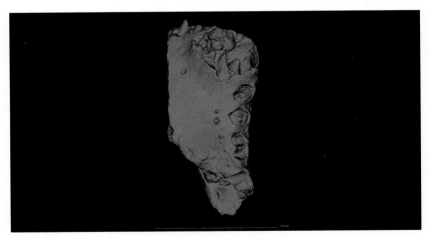

Video 12 Orthographic digital model of the right dentary from an adult *D. motherali*, DMNH 2013–07–0001. Video available at www.cambridge.org/drumheller.

Lateral to teeth d7 and d8, the lateral margin of the dentary is slightly concave for reception of the enlarged maxillary pseudocanines m4 and m5. In occlusal view, the anterior alveoli occur close to the lateral edge and shift medially going posteriorly. The medial surface of the ramus is open, exposing the long Meckelian groove that extends forward to the symphysial margin. The ventral and lateral surfaces of the dentary are smooth and ornamented with small, shallow pits and grooves. Neurovascular foramina along the lateral surface form a shallow groove below the alveolar margin. The right and left dentaries of DMNH 2013–07–1859 are nearly complete, missing only the posterior portions of the rami. In the smaller, complete left dentary of DMNH 2013–07–0240 (Figure 5C; Figure 17H), the posteriormost process is forked dorsoventrally, with a broad posteromedial process and short, narrow posterodorsal and posteroventral processes. The process overlaps the angular laterally before projecting medially to be overlapped by the angular and surangular. The posterior process fits together with the angular and surangular in a tongue and groove articulation so that an external mandibular fenestra is absent (Figure 20).

There are 21 dentary tooth positions preserved in the right dentary of DMNH 2013–07–1859 and 17 in the left. The alveoli increase in size from the first alveolus, which is the smallest, to d3 and d4, which are the largest. D3 and d4 are roughly equal in size, producing the dual pseudocanines as seen in the maxilla. The diameter is reduced at d6 and gradually increase in size through d15, decreasing thereafter.

4.13 Splenial

The right splenial is still attached to the right dentary of DMNH 2013–07–1859 but has shifted dorsally (Figure 17E–F). The splenial is long and triangular, narrowing anteriorly to extend to the level of the seventh dentary alveolus to participate in the mandibular symphysis. Posteriorly, it widens to the same level as the dentary and forms the medial half of the ventral surface of the mandible. There does appear to be a foramen for the intermandibularis oralis of the trigeminal nerve (CN V) on the anteriormost lateral surface of the splenial, but it is slightly obscured by the shifting of the bone.

4.14 Surangular

The right surangular of DMNH 2013–07–1859 is incomplete, representing only the posterodorsal corner of the element (Figure 2B). However, a similar sized individual (DMNH 2014–06–01) preserves a nearly complete posterior mandible (Figure 18A–D). The surangular forms the narrow dorsal margin of the posterior mandible. It has a nearly flat dorsal margin for the insertion of *M. adductor mandibulae externus* and the convex lateral surface is sculpted by round pits. It contributes to the lateral wall of the glenoid fossa. The ventral margin is thin with an inset groove for articulation with the angular. Anteriorly, this groove extends dorsolaterally for articulation with the posterodorsal process of the dentary. Anteromedially is a shallow groove for the articulation with

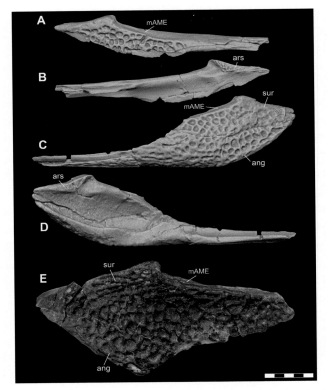

Figure 18 Lower jaw elements of subadult and adult individuals of
D. motherali. Orthographic image of 3D digital models of right surangular
DMNH 2014–06–01 in **A**, lateral, **B**, medial views; orthographic image of 3D
digital models of left surangular and angular DMNH 2014–06–01 in **C**, lateral,
D, medial views; right surangular and angular SMU 76810 in **E**, lateral view.
See text for anatomical abbreviations. Scale bar equals 5 cm.

the splenial. SMU 76810 preserves the posterodorsal corner of a right surangular (Figure 18E). It is heavily sculptured with large, tightly packed, oval- to
rectangular-shaped pits. As with DMNH 2014–06–01, the dorsal surface of the
nearly complete retroarticular process is caudoventrally angled in lateral view.

4.15 Angular

DMNH 2014–06–01 preserves a complete right and partial left angulars
(Figure 3). DMNH 2013–07–1859 also has a nearly complete left angular
and partial right (Figures 2 and 19A–D), while SMU 76810 is only the
posterioromost end of the element (Figure 18E). The angular forms the

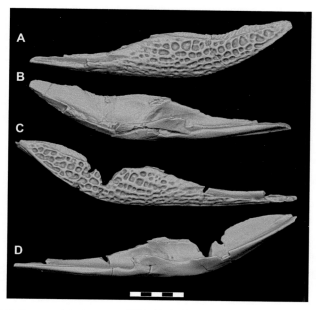

Figure 19 Orthographic image of 3D digital models of lower jaw elements of subadult individuals of *D. motherali*. Left angular DMNH 2013–07–1859 in **A**, lateral and **B**, medial views; right angular DMNH 2014–06–01 **C**, lateral, and **D**, medial views. See text for anatomical abbreviations. Scale bar equals 5 cm.

posteroventral portion of the posterior mandible. Its lateral surface is almost entirely sculpted by round pits that become smaller anteriorly and form shallow grooves similar to those seen on the dentary. The posterodorsal margin is rugose and slightly deflected for articulation in the inset groove of the surangular. Posteriorly, the angular narrows to extend to the posterior end of the retroarticular process. An inset groove occurs on the anterolateral process of the angular for articulation with the posteroventral process of the dentary (Figure 20A, B). When articulated, the tongue and groove articulation between the angular, surangular, and posteroventral process of the dentary result in no external mandibular fenestra. This is remarkably similar to that of *P. newmani* (SMU 76601) (Figure 20C). On the anteroventral margin of the angular is a shallow groove for contact with the splenial. A shallow fossa occurs on the posteromedial surface, ventral to the position of the retroarticular process, for the attachment of the *M. pterygoideus ventralis*. Medially, a trough-like concavity of the angular forms the floor of the mandibular adductor fossa posteriorly and the Meckelian canal anteriorly.

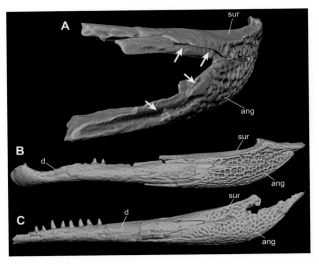

Figure 20 Orthographic image of 3D digital models of left surangular and angular of *D. motherali*, DMNH 2014–06–01 articulated to show the groove (arrows) for the insertion of the posterior process of the dentary in **A**, oblique view. Orthographic image of 3D digital reconstruction of the lower jaw with the left dentary of DMNH 2013–07–0240 in **B**, lateral view. Posteriormost process of dentary missing from 3D model indicated with hatched lines. Orthographic image of 3D digital reconstruction *P. newmani* lower jaw SMU 76601 (reflected) **C**, lateral view. See text for anatomical abbreviations. Not to scale.

4.16 Articular

DMNH 2013–07–1859 does not preserve either articular. As with the surangular, the similar sized DMNH 2014–06–01 does preserve complete left and right elements (Figure 21). The articular extends anteroposteriorly the length of the corresponding retroarticular process of the surangular. The glenoid fossa is figure-eight-shaped, composed of two asymmetric depressions bisected by a prominent ridge. A prominent dorsal ridge marks the posterior border of the glenoid fossa. The surangular does not participate in the glenoid surface. The descending ramus of the articular is triangular in cross section, tapering ventrally to a thin lamina that passes along the medial surface of the surangular. Anteriorly, this lamina is directed medially following the prominent ridge that divides the glenoid fossa. In lateral view, the sutural surface for the surangular is slightly concave and tapers posteriorly to become triangular-shaped. In dorsal view, the retroarticular process is elongate and paddle-shaped, with a gently concave dorsal surface. The retroarticular process is oriented horizontally

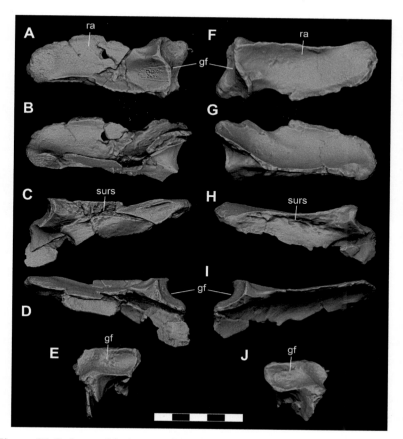

Figure 21 Orthographic image of 3D digital models of articulars of subadult individuals of *D. motherali*, DMNH 2014–06–01. Left articular in **A**, dorsal, **B**, ventral, **C**, medial, **D**, lateral, and **E**, anterior views; right articular in **F**, dorsal, **G**, ventral, **H**, lateral, **I**, medial, and **J**, anterior views. See text for anatomical abbreviations. Scale bar equals 5 cm.

relative to the long axis of the mandible. There is no indication of a foramen aëreum at the anteromedial edge of the retroarticular process.

4.17 Dentition

The teeth of *D. motherali* are like most crocodyliforms in being conical and circular in cross section (Figure 22). The large teeth referable to the holotype (DMNH 2013–07–0001) are upwards of 55 mm in length and 20 mm in width, and exhibit longitudinal ridges that terminate before the apex and with carinae that lack denticles. The enamel ridges and carinae become less prominent with

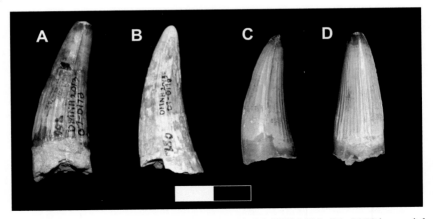

Figure 22 Isolated teeth of *D. motherali*. **A, B,** DMNH 2013–07–0178 in mesial or distal views WM 2019–15 Gb in **C**, mesial or distal and **D**, lingual views. Scale bar equals 2 cm.

increasing tooth size. All the teeth exhibit a labial curvature that is more pronounced in the larger crowns. Several juvenile and subadult specimens have complete teeth, as well as replacement teeth, preserved in situ, although much of the subadult material tend to be missing their crowns. The teeth of the juveniles preserve closely spaced, nonanastomosing ridges that terminate shortly before the apex. The carinae are oriented mesiodistally and lack denticles.

4.18 Pathologies

Several elements attributable to *Deltasuchus* exhibit pathologies. The left maxilla of DMNH 2013–07–1859 has a deep depression on the dorsal surface of the rostrum, situated near the medial margin and roughly parallel with the m10 alveolus (Figure 9A, D, E). In dorsal view, this feature is 7.77 mm in maximum width, 22.87 mm in maximum length, and 6.34 mm in maximum depth. The margins of this structure are smooth in the interior of the depression, but the sculpturing that characterizes the rest of the maxilla is present, if less defined, at the margins of the structure. The edges of the indentation, as well as the transition from unsculptured to sculptured bone, is smooth with no visible fracturing present in this aspect. In medial view, a wedge of bone is present directly beneath the depression, within the nasal passage. It connects with the roof of the nasal passage, suggesting that it represents the dorsal maxillary bone displaced by the overlying, indented feature. No obvious fracturing is visible, and the wedge is narrowest anteriorly, widening posteriorly. The texture is somewhat fibrous and irregular, but the entire element exhibits minor flattening and deformation, which is particularly visible within the nasal passage. The maxilla is broken just anterior to this pathology, and while the element has been glued back together, portions of the bone are missing from both dorsal and medial margins, further affecting the modified region of the element.

Taken together, this structure appears to be partially healed, localized impact or crushing damage consistent with a bite mark. In modern crocodylians, intraspecific competition often is expressed as bites to the rostra, limbs, and bases of tails, making injuries in these regions fairly common (Webb et al., 1982). Similar placement of bite marks in other crocodyliform taxa have been used as evidence of intraspecific competition in other fossil groups (Buffetaut, 1983; Williamson, 1996; Avilla et al., 2004; Katsura, 2004; Mackness et al., 2010; Vasconcellos and Carvalho, 2010). However, the degree of partial healing, partnered with the lack of other serial bite marks or more diagnostic features, makes this specific interpretation somewhat speculative. At best, we can say that the injury is consistent with a deep, if partially healed, puncture (sensu Binford, 1981). Given the faunal composition of the site, neither medium to large crocodyliforms nor theropod dinosaurs can be completely excluded as potential trace makers.

The fragment of dentary within associated DMNH 2014–06–01 elements also exhibits a bite mark. This one, by comparison, still exhibits crushing damage from the initial impact. It does not fully pierce the cortical bone, making it a pit according to Binford's (1981) bite mark classification scheme. The pit is teardrop-shaped, being 6.34 mm in maximum width and 8.64 mm in maximum length, and is U-shaped in cross section. The mark morphology lacks an obvious bisection, which would have been diagnostic of Crocodyliformes (Njau and Blumenschine, 2006; Drumheller and Brochu, 2014, 2016; Drumheller et al., in press), but it is consistent with a large, conical tooth, making a crocodyliform the most likely trace maker from the AAS assemblage. Theropod dinosaurs exhibit ziphodont dentition, which would result in a narrower, more V-shaped mark in cross section (D'Amore and Blumenschine, 2009), and none of the mammals or other toothed vertebrates present at the site would have been large enough to have left this trace (Noto et al., 2012). Again, however, it is not possible to make a more specific association with any one of the several species present at the site, but intraspecific competition is a possible source of the injury.

5 Phylogenetic Analysis

5.1 Methods

Previous analyses have recovered *Deltasuchus motherali* forming a clade with *Paluxysuchus newmani*, basal to Goniopholididae and other neosuchians

(Adams et al., 2017; Adams, 2019; Kuzmin et al., 2019; Noto et al., 2019). In order to test the evolutionary relationships of *D. motherali* and *P. newmani* within Neosuchia, a set of phylogenetic analyses were conducted using revised character codings for *D. motherali* that include new information derived from those elements of adult and subadult individuals that do not demonstrate any ontogenetic variability. The first phylogenetic analyses incorporated an updated version of the character-taxon matrix of Turner (2015) which was modified in later studies by Adams et al. (2017) and Adams (2019) (see Appendix S1 for character list). The original dataset by Turner (2015) initially contained 318 active characters and 101 taxa. Following Kuzmin et al. (2019), the analysis presented here is revised and expanded to include a crocodyliform taxon (TMM 42536–2) from the Upper Cretaceous (early Campanian) Aguja Formation of Brewster County, Texas for a total of 321 osteological characters and 107 crocodylomorpha taxa. TMM 42536–2 was scored for 130 of the 321 characters based on Lehman et al. (2019).

The second matrix for phylogenetic analysis employed the merged character-taxon matrix originally published by Young et al. (2016), which was then subsequently revised and expanded by Ristevski et al. (2018), who referred to it as the Hastings + Young matrix (H+Y matrix) (Appendix S2). The dataset by Ristevski et al. (2018) initially contained 387 morphological characters and 137 taxa but was recently expanded by Lehman et al. (2019) to include TMM 42536–2, the crocodyliform from the Upper Cretaceous (early Campanian) Aguja Formation of Brewster County, Texas. *D. motherali* and *P. newmani* were added to the H+Y matrix for a total of 387 characters and 140 taxa.

The analyses were conducted using TNT v. 1.5 (Goloboff et al., 2008; Goloboff and Catalano, 2016). A heuristic tree search strategy was conducted performing 1,000 replicates of Wagner trees (using random addition sequences) followed by TBR branch swapping (holding 10 trees per replicate). The Turner matrix treated 34 characters as ordered, while 19 characters were treated as ordered for the H+Y matrix. All other characters were equally weighted. The best trees obtained at the end of the replicates were subjected to a final round of TBR branch swapping. Zero-length branches were collapsed if they lack support under any of the most parsimonious reconstructions. Character support of the nodes present in the most parsimonious reconstructions was calculated using bootstrap analysis of 1,000 replicates and the BREMER.RUN script for Bremer support (Bremer 1988, 1994). The topologies obtained during the bootstrap replicates are summarized using frequency differences (Groups present/Contradicted, GC), following Goloboff et al. (2003). Phylogenetic nomenclatural and clade definitions follow that of Brochu et al. (2009) and Turner (2015).

5.2 Results

The first analysis using the revised matrix of Turner (2015) resulted in 4 most parsimonious trees (MPTs), with a length of 1,728 steps, a consistency index (CI) of 0.233, and a retention index (RI) of 0.666 (Figure 23). Within Neosuchia, the strict consensus topology resulted in several differences from those found in Adams et al. (2017), Adams (2019), and Noto et al. (2019). As with previous results, *D. motherali* was recovered inside Neosuchia in all four MPTs as the sister taxon to *P. newmani*. This clade is supported by two unambiguous synapomorphies: absence of a mandibular fenestra (character 74.1); postorbital in contact with the ectopterygoid (character 143.0). Unlike previous analyses, Goniopholididae is positioned more basal to the clade containing *Paluxysuchus* + *Deltasuchus* (hereafter referred to as Paluxysuchidae). TMM 42536–2 was recovered basal to Paluxysuchidae in all four MPTs. A TMM 42536–2 + Paluxysuchidae + Thalattosuchia + Tethysuchia clade is defined by three synapomorphies: postorbital-jugal contact anterior to jugal (character 16.0); jugal does not exceed the anterior margin of orbit (character 121.0); base of postorbital process of jugal directed dorsally (character 141.1). Differences were also recovered within Eusuchia with the overall typology being more resolved, with fewer polytomies occurring in Crocodylia and Paralligatoridae. A monophyletic clade containing *Wannchampsus* sp. (formerly known as the Glen Rose Form sensu Adams, 2014) and *Wannchampsus kirpachi* was recovered as a sister group to Atoposauridae, and not as a paralligatorid, supporting previous hypothesis of closer affinities between atoposaurids and paralligatorids (Turner, 2015; Schwartz et al., 2017; Kuzmin et al., 2019). The addition of *Kansajsuchus extensus*, *Turanosuchus aralensis*, and the Dzharakuduk paralligatorid resulted in Paralligatoridae showing a similar topology to that of Kuzmin et al. (2019), with *Tarsomordeo winkleri* and *Rugosuchus nonganensis* being pulled down to more basal positions. This clade is defined by six synapomorphies: midline ridge on the dorsal surface of frontal and parietal (character 21.1); unsculptured "lobe" on the posterodorsal corner of the squamosal (character 34.1); basisphenoid exposed on the ventral surface of braincase (character 55.0); external nares divided by a septum (character 65.0); sharp ridge along the lateral surface of the angular (character 218.2); width of posterior half of the axial neural spines are narrow (character 257.1).

The results of the second analysis utilizing the H+Y matrix recovered 130 MPTs with 1,285 steps (CI = 0.406, RI = 0.843 Figure 24). Except for the inclusion of *D. motherali* and *P. newmani*, the overall typology of the strict

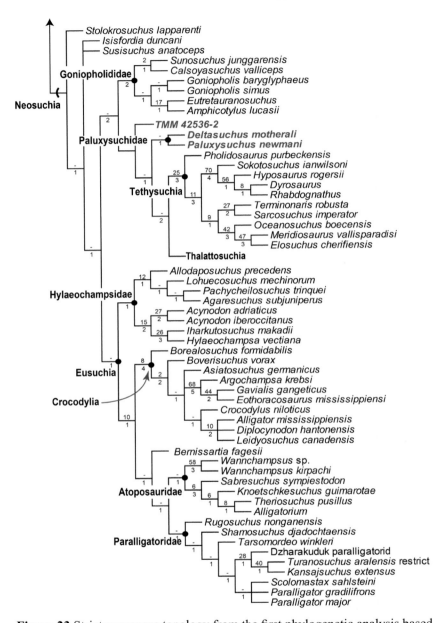

Figure 23 Strict consensus topology from the first phylogenetic analysis based on the Turner matrix of 107 taxa and 321 characters. Four equally MPTs of 1,728 steps (CI = 0.233 and RI = 0.666) were obtained from a cladistic analysis. Paluxysuchidae are indicated in red. Numbers at each node indicate bootstrap GC values (top number) and Bremer support values (bottom number). Semilunar hash marks indicate stem-based definition and solid circles indicate a node-based taxon.

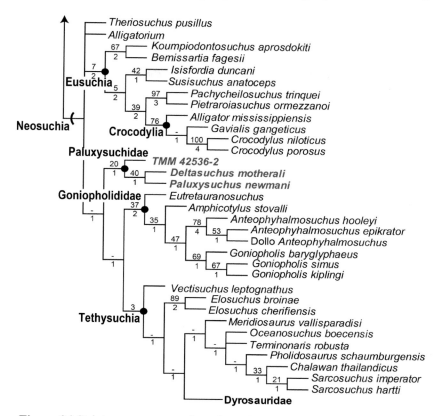

Figure 24 Strict consensus topology from the second phylogenetic analysis based on the H+Y matrix using 140 taxa and 387 characters, resulting in 130 equally MPTs of 1,285 steps (CI = 0.406 and RI = 0.843). Paluxysuchidae are indicated in red. Numbers at each node indicate bootstrap GC values (top number) and Bremer support values (bottom number). Semilunar hash marks indicate stem-based definition and solid circles indicate a node-based taxon.

consensus tree does not vary from that of Ristevski et al. (2018). *D. motherali* and *P. newmani* form a monophyletic clade (Paluxysuchidae) basal to Goniopholididae. *D. motherali* + *P. newmani* is defined by one synapomorphy: absence of a mandibular fenestra (character 216.1). The most significant difference between Ristevski et al. (2018) and this current analysis is the placement of TMM 42536–2 in a basal position within Paluxysuchidae in all 130 MPTs. *Paluxysuchus* + *Deltasuchus* + TMM 42536–2 is defined by five synapomorphies: lacrimal reaches the anterodorsal and anteroventral margins of the orbit (characters 120.1 and 122.1); width of anterior process and posterior process of the jugal are subequal in size (character 131.0); surangular extends to posterior end of retroarticular process (character 229.1); surface of the retroarticular process is paddle shaped (character 236.2).

6 Discussion

6.1 Relationship of Paluxysuchidae and TMM 42536–2

Overall, the results of the phylogenetic analyses confirm the inclusion of both *Paluxysuchus newmani* and *Deltasuchus motherali* into a common clade, Paluxysuchidae, clade nov. Of particular interest is the placement of TMM 42536–2 within Paluxysuchidae in the H+Y matrix, a result of clear similarities; triangular form of the lacrimal and its involvement in the orbital margin, jugal not exceeding the anterior margin of the orbit, the anterior and posterior ramus of the jugal being subequal, ventrally expanded medial condyle of the quadrate, and a paddle-shaped retroarticular process projecting dorsally. Lehman et al. (2019) had recovered TMM 42536–2 in a basal position within Goniopholididae, referring to it as the "Aguja goniopholidid." In their only comparisons, they distinguished it from the only other purported goniopholidid crocodyliform known from Upper Cretaceous of North America, *Denazinosuchus kirtlandicus* from New Mexico, based on the presence of a vestigial maxillary fossa, a small mandibular fenestra, and large suborbital fenestra. The shallow maxillary fossa, a character typically used to recognize goniopholidids (character 48.1 in the H+Y matrix), is described as "vestigial (or rudimentary)" (Lehman et al., 2019: p. 302) and could possibly be analogous to the shallow groove found on the posterior maxillary process of *D. motherali*, which can be attributed to surface sculpturing or potentially the confluence of several maxillary neurovascular foramina on the lateral margin of the maxilla (Adams et al., 2017). It would require five extra steps to retrieve TMM 42536–2 within Goniopholididae in the H+Y matrix as found in Lehman et al. (2019) and three extra steps in the Turner matrix.

The position of TMM 42536–2 in the Turner matrix places it outside Paluxysuchidae, positioned as a separate, but closely related taxon. It requires only one extra step to retrieve it within that clade. The differences in the two analyses may be a result of the fragmentary nature of TMM 42536–2, with the specimen lacking diagnostic elements (Lehman et al., 2019). More complete and better-preserved specimens may help resolve the two alternative phylogenies presented here. Additionally, the close connection between TMM 42536–2, *D. motherali*, and *P. newmani* strengthens the argument for the early emergence of an endemic crocodyliform assemblage at the beginning of the Early Cretaceous and implies that the transition from the Early to Late Cretaceous may have been more gradual for crocodyliforms (Adams et al., 2015, 2017).

6.2 Relationships of Paluxysuchidae and Closely Related Groups

The clade TMM + Paluxysuchidae was recovered as either basal to Goniopholididae + Tethysuchia in the H+Y matrix or to Tethysuchia + Thalattosuchia in the Turner. In the H+Y matrix, it would require only one extra step to position Paluxysuchidae basal to Tethysuchia, while taking four extra steps in the Turner matrix to place it basal to Goniopholididae. The close relationship between Goniopholididae + TMM + Paluxysuchidae + Tethysuchia + Thalattosuchia from the Turner analysis is characterized by eight synapomorphies: nasals do not contribute to the narial border (12.1); anterior edge of choanae situated between the suborbital fenestrae (43.0); dorsal osteoderms have a well-developed process located anterolaterally on dorsal parasagittal osteoderms (95.2); osteoderms without longitudinal keels (100.1); postzygapophyses of axis are poorly developed (152.1); supraoccipital not exposed on the skull roof (170.0); projection of anterior alveoli of the dentary are weakly procumbent (261.1); no midline crest on the basioccipital plate below the occipital condyle (294.0). A Paluxysuchidae + Tethysuchia + Thalattosuchia clade is supported by only one transformation: splenial extensively involved in symphysis in ventral view (76.2). However, there are additional characters shared between these groups that help to diagnose Paluxysuchidae: the posterior ramus of the jugal beneath the infratemporal fenestra is rod-shaped (17.1; shared with pholidosaurids); retroarticular process facing dorsally and paddle shaped (70.3; shared with pholidosaurids); jugal does not exceed the anterior margin of the orbit (121.1; shared with tethysuchians). The results of this analysis support that these characters are not dependent of a tubular-snouted longirostrine condition, as proposed by Pol and Gasparini (2009), but are more likely a homoplastic condition (Adams, 2013).

The H+Y analysis resulted in four synapomorphies for a TMM + Paluxysuchidae + Goniopholididae + Tethysuchia clade: the presence of large depressions on the ventral surface of the premaxilla posterior to either the P1–P2 for reception of the d1 teeth (24.1); the anterior and anterolateral margins of premaxilla are subvertical, and extend ventrally to the rest of the element (26.1); dorsal surface of lacrimal extend laterally over the orbit (56.1); median process of the frontal extends anterior to the tip of the prefrontal (94.0). Presacral dorsal armor considerably wider than long (372.1). A Goniopholididae + Tethysuchia clade is associated with two synapomorphies: the presence of an enlarged foramina and associated fossae on the lateral margin of the posterior maxillae and/or the anterior process of the jugal (47.1); quadrate and squamosal do not laterally enclose the cranioquadrate canal, and it is laterally exposed (204.2). In either analysis, the close affinity of these clades is not surprising as they share a number of morphological features (see Section 5.2, Results) and strengthens the existence of a neosuchian lineage, comprised of mesorostrine platyrostral forms and tubular-snouted longirostrine taxa, that diverges from the lineage that led to Eusuchia (Andrade et al., 2011).

6.3 Niche Partitioning and Ontogeny

In crocodyliforms, snout shape (e.g. Brochu, 2001; Drumheller and Wilberg, 2020) and tooth shape (e.g. D'Amore et al., 2019) have been associated with diet and feeding strategy. Slender snouts and thin, conical teeth are associated with predation on smaller- and softer-bodied prey (i.e. Iordansky, 1973; Langston, 1973; Busby, 1995; McHenry et al., 2006; Gignac and O'Brien, 2016; Drumheller and Wilberg, 2020) while blunt snouts and anvil-like teeth are specializations associated with durophagy (e.g. Brochu, 2001; D'Amore et al., 2019; Drumheller and Wilberg, 2020). Broad U- or V-shaped snouts, especially when partnered with sturdy, conical teeth, are found in taxa that exhibit a more generalist diet, including full-blown macropredation (e.g. Brochu, 2001; Drumheller and Wilberg, 2020). Morphometric analyses of snout shape have largely supported these divisions within the crown group (Pierce et al., 2008; Sadleir and Makovicky, 2008), but inclusion of more distantly related fossil taxa further subdivides Crocodyliformes into two functionally separate slender-snouted and generalist ecomorphotypes (Wilberg, 2017; Drumheller and Wilberg, 2020).

Adult *Deltasuchus* specimens exhibit the broadly triangular snout shape that is associated with a generalist diet, but not macropredation, i.e. active predation on taxa that are larger than the predator (Drumheller and Wilberg, 2020) (Figure 25A, B; Video 13A, 13B, 14A, 14B). Partnered with stout, conical teeth (Adams et al., 2017; D'Amore et al., 2019), these animals were well-adapted to feeding on the wide diversity of prey items present in the AAS, an interpretation that is further bolstered by the presence of bite marks attributable to this taxon on dinosaurian and turtle material at the site (Noto et al., 2012; Adrian et al., 2019). However, the diet and feeding strategy can shift through age and ontogeny, and interpretations applicable to adults may not reflect the paleoecology of juveniles.

Crocodyliforms increase in size by several orders of magnitude during growth and development, which has a direct effect on the prey items available across the range of body sizes seen within a single species (Gignac and O'Brien, 2016; Gignac et al., 2019). In modern groups, very young individuals are commonly observed to eat a diversity of small-bodied prey, including arthropods, fish, and amphibians (e.g. McIlhenny, 1935; Ross and Magnusson, 1989),

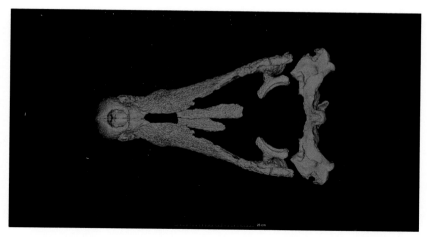

Video 13 Composite, orthographic digital model of an adult *D. motherali* cranium and mandible. Video available at www.cambridge.org/drumheller.

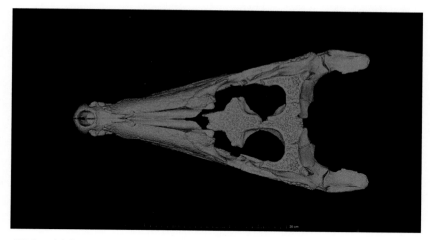

Video 14 Composite, orthographic digital model of a subadult *D. motherali* cranium and mandible. Video available at www.cambridge.org/drumheller.

and it is only into adulthood that the blunter-snouted taxa shift away from compliant prey items and take on the generalist diets more commonly associated with their respective rostral and dental morphologies (Drumheller and Wilberg, 2020).

In *Deltasuchus*, the juvenile material described in this study does exhibit changes in rostral and dental morphology consistent with shifting niche

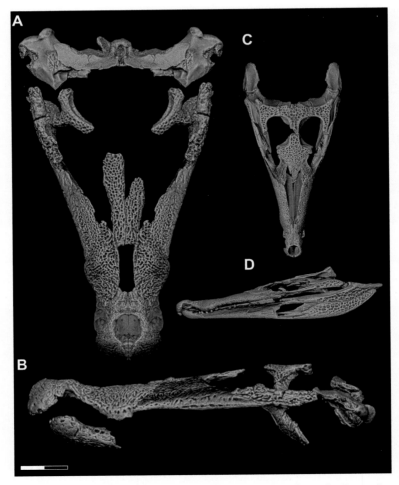

Figure 25 Orthographic image of 3D digital reconstruction of a *D. motherali* adult skull in **A**, dorsal, **B**, lateral views and a subadult skull in **C**, dorsal, **D**, lateral views. Scale bar equals 10 cm.

occupation during ontogeny. The angle of the snout widens noticeably between juveniles and subadults. Measurements of the smallest specimen (DMNH 2013–07–0228) suggest that when articulated, the dentaries would have diverged from one another at an angle of about 20°, while the most complete subadult (DMNH 2013–07–1859) had a comparatively wider snout, with dentaries diverging from one another at 30° (Figure 25C, D). The largest adult specimen is the holotype (DMNH 2013–07–0001), but the dentaries associated with this individual are very fragmentary making direct measurement difficult.

Extrapolating an angle of the snout from the rostral reconstruction yields a similar angle to the subadult, roughly at 30° (Figure 25).

Dentition also shifts through ontogeny. Size heterodonty remains fairly consistent, with the paired pseudocanines of both the dentaries and maxillae of all size classes having roughly twice the alveolar diameters of their immediate neighbors across all specimens and size classifications (Table 1). Not all jaw elements contain complete teeth, but in those that do, there is a trend in overall shape with increased size. The teeth from the smallest juveniles (e.g. DMNH 2013–07–0228, DMNH 2013–07–0240, DMNH 2013–07–0312, DMNH 2013–07–0322) are noticeably more slender than those in the subadults (e.g. DMNH 2013–07–1859) and adults (e.g. DMNH 2013–07–0001).

These two changes during ontogeny suggest a shift in the niche occupied by different size and age *Deltasuchus*. In the larger context of the site, these changes make sense given the other crocodyliform taxa present in the site. In addition to common *Deltasuchus* material, the AAS has also yielded crocodyliform fossils from at least three other species representing two distinct ecomorphotypes. *Scolomastax* was a small-bodied crocodyliform with a short snout and significant size (and probable shape) heterodonty, suggesting that it specialized in hard food items. It is possible that this species was an omnivore or maybe even a herbivore (Noto et al., 2019). *Terminonaris* and *Woodbinesuchus*, in comparison, were both large-bodied, slender-snouted and -toothed species (Lee, 1997a; Adams et al., 2011) that would have specialized on smaller-bodied, more compliant prey (Drumheller and Wilberg, 2020). Between these two species, further niche partitioning is suggested based on relative snout length and environmental preference. *Terminonaris* had a comparatively longer snout, and is found in coastal, nearshore environments. In comparison, *Woodbinesuchus* exhibited a slightly less longirostrine condition, and seems to be more restricted to freshwater to brackish environments (Jouve and Jalil, 2020; Jouve et al., in press). The geology of the AAS is complex, including freshwater, brackish, saltwater, and terrestrial input (Adams et al., 2017; Noto et al., 2019), and therefore would have preserved organisms in direct association that may not have overlapped significantly in life.

The ontogenetic trajectory of *Deltasuchus* would have differentiated members of this group from other crocodyliforms in its immediate vicinity throughout life. The more gracile morphology of the juveniles would not lend itself well to consuming hard prey items, but occupying this niche potentially would have brought them into direct competition with *Scolomastax*. However, even the youngest specimens are not as longirostrine as sympatric *Terminonaris* and *Woodbinesuchus*, suggesting that these groups were also responding to pressures to partition niches. With age, increased size and snout robusticity would

Table 1 Measurements of labiolingual tooth diameters in mms. Teeth are listed by anatomical position (p = premaxilla; m = maxilla; d = dentary) and numbered mesially to distally. Blank = no alveolus preserved for this position on this specimen; – = alveolus is preserved, but is too damaged or incomplete to reliably measure; ? = uncertainty with regards to the dental count; * = tooth/replacement tooth present

Tooth	Specimen#								
	DMNH 2013–07–0001 left	DMNH 2013–07–0001 right	DMNH 2013–07–0297	DMNH 2013–07–1859 left	DMNH 2013–07–1859 right	DMNH 2013–07–0079	DMNH 2013–07–0239	DMNH 2013–07–1888	DMNH 2013–07–0219
p1	–	9.7	6.87	6.15					
p2	–	20.99	–	7.78*					
p3	–	22.28	11.51	–					
p4	–	12.42	7.81	–					
m1	11.54			5.57		6.05			4.04
m2	12.44	15.45		6.25	–	6.72			4.57
m3	17.45	15.89		6.5	–*	8.04*			4.67
m4	**24.53**	**24.86**		**11.66**	–	**12***			**8.26**
m5	**24.03**	**24.94**		**10.62***	**11.58***	**12.82***			**8.94**
m6	12.9	–		7.1	8.45	7.84			4.67
m7	14.09	–		6.81	7.43	7.93			4.29
m8	15.18	15.14		6.48	7.06	6.74			3.79
m9	15.23	17.44		6.25	–	6.61			5.18

m10	–	18.19	6.61	7.53	6.75
m11	19.26	19.22	6.61	–	7.67
m12	18.32	18.15	–	–	
m13	17.62	18.73		8.26	
m14	16.99	17.55		–	
m15	15.63	15.9		–	
m16	14.03	15.3		–	
m17		13.82		–	
m18		11.63			
m19		11.06			
m20		8.67			
m21		7.01			
m22		6.91			
d1	11.78*	11.58*	3.64	5.51*	–	–
d2	10.21*	11.44*	–	5.85	7.3	
d3	**22.13***	**–***	**8.63**	**11.43**	**12.54**	–
d4	**23.48***	**23.45***	**7.25**	**11.65**	**13.13**	**10.5**
d5	12.66	13.32*	4.47*	6.64*	7.79	7.68
d6	13.54	13.31	5.44*	6.49	7.19	7.1
d7	11.35	11.38	5.04*	5.44*	6.34	6.21
d8	10.24	–	–	5.94*	5.79	–
d9			5.60*	6.36*	5.59	–

Table 1 (cont.)

Tooth	Specimen#				
d10	7.4	6.31*	—	6.41	7.87?
d11	—	—*		7.14	6.99?*
d12	8.13*	6.72*		7.49	6.64?
d13	7.17*	6.54		8.34	6.09?
d14	6.83	6.56*		7.65	5.80?
d15	6.22*	—*		7.45	6.89?
d16	4.11	—*		6.72	5.73?
d17	4.78	—*		6.36	—?
d18	—	—*			
d19	—	—*			
d20	—	—*			
d21	—	—*			
d22		—			
d23		—			
d24					

Tooth	DMNH 2013–07–1984	DMNH 2013–07–0218	DMNH 2013–07–0322	DMNH 2013–07–0802	DMNH 2013–07–0240	DMNH 2013–07–0228	DMNH 2013–07–0312
d1	3.73*		4.09*	3.39	2.71	2.08	
d2	3.79		3.81	–	–	2.16	
d3	**8.14***		**7.59**	–	**5.27***	**4.1**	
d4	**8.04***		**8.52**	**6.62**	**5.00***	**3.97**	
d5	3.81	–	4.1*	3.91	3.64	1.97	
d6	3.82*	–	3.53	3.41		2.05	
d7	3.24	2.37	3.44	2.78		1.96	
d8	2.87	2.7	2.93	2.33	1.78	1.96	
d9	3.49	3.37	3.52		1.62	2.26	
d10	3.71	4.74	3.34		2.19	1.91*	
d11	–*	5.89	–		2.82	2.42	–?
d12	–	6.22	4.74*		2.7	2.86*	3.33?
d13		5.79	5.10*			2.24	3.81?*
d14		4.88	4.63*			2.32*	4.04?*
d15		5.63	4.26		3.75*	3	3.59?*
d16		5.39	3.92		3.79*	2.21*	3.89?*
d17			3.78		2.83*	2.06	3.73?*

(cont.)

Tooth	Specimen#			
d18	3.78	–	2.20*	2.88?*
d19	3.33	–	–	2.84?*
d20	2.75*	–		2.57?*
d21		–		2.36?*
d22		–		–?
d23				–?
d24				–?

have moved this taxon into an ecomorphotype that allowed predation on larger and hard-bodied prey items (Drumheller and Wilberg, 2020), an ecological interpretation that is further bolstered by the presence of bite marks attributable to midsized *Deltasuchus* on turtles and dinosaurs known from the AAS (Noto et al., 2012; Adrian et al., 2019). This shifted subadults and adults into a niche that was not occupied by the other large-bodied taxa in this ecosystem, which were all specializing in smaller, more compliant prey (Drumheller and Wilberg, 2020).

Similar examples of niche partitioning, as reflected by disparate snout and tooth morphologies, are known in other fossil localities with multiple species of sympatric crocodyliforms (e.g. Salas-Gismondi et al., 2015), but much of this previous focus has fallen on the morphology of the adult members of these clades. This is most likely being driven by availability of specimens, but in modern settings, dietary pressures seem to be exceptionally high on the youngest crocodylians, suggesting that selection for or against juvenile morphotypes is particularly important with regards to survival into adulthood and sexual maturity (e.g. Gignac and O'Brien, 2016; Gignac and Santana, 2016). The ontogenetic trajectory of increasing snout width and robusticity with growth and development, as observed in *Deltasuchus*, is expected, because it mirrors similar patterns observed qualitatively and quantitatively across all modern crocodylians (Drumheller et al., 2016; Iijima, 2017). The dietary implications of slender, more gracile juvenile snouts supporting slender dentition transitioning into wider, more robust, sturdier-toothed adult morphologies has been particularly well-explored across nonalligatoroid crocodylians, drawing from both isotopic (Radloff et al., 2012) and geometric morphometric evidence (Iijima, 2017). The availability of multiple specimens attributable to *Deltasuchus* in the AAS system provides a rare opportunity to discuss how juveniles fit into the greater pattern of niche occupation and partitioning among crocodyliforms in this site.

7 Conclusions

The fossil record of terrestrial and freshwater vertebrates, including crocodyliforms, is often highly fragmentary and taphonomically biased. Our knowledge of many taxa is based on single, incomplete specimens. When they are available, ontogenetic series across species provide invaluable information on intraspecific variation, critical to understanding character development in phylogenetic contexts and paleoecological niche shifts during growth and development. An expanded, ontogenetic series attributable to *Deltasuchus* yields paleoecological information regarding dietary shifts across age and size classifications, illuminating how juveniles fit into the trophic hierarchies and specializations of a crocodyliform-rich ecosystem. This material also provides further support of an endemic group of mid-Cretaceous neosuchians, currently best known from the subcontinent of Appalachia. Herein named Paluxysuchidae, clade nov., this clade encompasses *Paluxysuchus* (Adams, 2013), *Deltasuchus* (Adams et al., 2017), and potentially TMM 42536–2 from the Upper Cretaceous (early Campanian) of Texas, and represents a mid-Cretaceous radiation of crocodyliforms that coincided with the opening of the Western Interior Seaway.

References

Adams, T. L. 2013. A new neosuchian crocodyliform from the Lower Cretaceous (Late Aptian) Twin Mountains Formation of north-central Texas. *Journal Vertebrate Paleontology* 33: 85–101.

Adams, T. L. 2014. Small crocodyliform from the Lower Cretaceous (Late Aptian) of Central Texas and its systematic relationship to the evolution of Eusuchia. *Journal of Paleontology* 88: 1031–49.

Adams, T. L. 2019. Small terrestrial crocodyliform from the Lower Cretaceous (late Aptian) of central Texas and its implications on the paleoecology of the Proctor Lake Dinosaur locality. *Journal of Vertebrate Paleontology* 39(3). https://doi.org/10.1080/02724634.2019.1623226

Adams, R. L. and J. P. Carr. 2010. Regional depositional systems of the Woodbine, Eagle Ford, and Tuscaloosa of the U.S. Gulf Coast. *Gulf Coast Association of Geological Societies Transactions* 60: 3–27.

Adams, T. L., M. J. Polcyn, O. Mateus, D. A. Winkler, and L. L. Jacobs. 2011. First occurrence of the long-snouted crocodyliform *Terminonaris* (Pholidosauridae) from the Woodbine Formation (Cenomanian) of Texas. *Journal of Vertebrate Paleontology* 31: 712–16.

Adams, T. L., C. R. Noto, and S. Drumheller. 2015. The crocodyliform diversity of the Woodbine Formation (Cenomanian) of Texas and the transition from Early to mid- Cretaceous ecosystems. *Journal of Vertebrate Paleontology, SVP Program and Abstracts Book* 2015: 77.

Adams, T. L., C. R. Noto, and S. Drumheller. 2017. A large neosuchian crocodyliform from the Upper Cretaceous (Cenomanian) Woodbine Formation of North Texas. *Journal of Vertebrate Paleontology*. https://doi.org/10.1080/02724634.2017.1349776

Adrian, B., H. F. Smith, C. R. Noto, and A. Grossman. 2019. A new baenid, "Trinitichelys" maini sp. nov., and other fossil turtles from the Upper Cretaceous Arlington Archosaur Site (Woodbine Formation, Cenomanian), Texas, USA. *Palaeotologia Electronica*. https://doi.org/10.26879/1001

Ambrose, W. A., T. F. Hentz, F. Bonaffé, et al. 2009. Sequence stratigraphic controls on complex reservoir architecture of highstand fluvial-dominated deltaic and lowstand valley-fill deposits in the Upper Cretaceous (Cenomanian) Woodbine Group, East Texas field: Regional and local perspectives. *American Association of Petroleum Geologists Bulletin* 93: 231–69.

Andrade, M. B., R. Edmonds, M. J. Benton, and R. Schouten. 2011. A new Berriasian species of *Goniopholis* (Mesoeucrocodylia, Neosuchia) from

England, and a review of the genus. *Zoological Journal of the Linnean Society* 163: S66–108.

Avilla, L. S., Fernandes, R., and Ramos, D. F. B. 2004, Bite marks on a crocodylomorph from the Upper Cretaceous of Brazil: evidence of social behavior? *Journal of Vertebrate Paleontology* 24: 971–3.

Bailleul, A. M., J. B. Scannella, J. R. Horner, and D. C. Evans. 2016. Fusion patterns in the skulls of modern archosaurs reveal that sutures are ambiguous maturity indicators for the Dinosauria. *PLoS ONE* 11(2): e0147687. https://doi.org/10.1371/journal.pone.0147687

Barnes, V. E., J. H. McGowen, C. V. Proctor, et al. 1972. Geologic Atlas of Texas, Dallas Sheet, 1:250,000 scale. University of Texas at Austin, Bureau of Economic Geology.

Bellairs, A. 1969. *The Life of Reptiles*. Vol. II. London: Weidenfeld and Nicolson, p. 307.

Benton, M. J. and J. M. Clark. 1988. Archosaur phylogeny and the relationships of the Crocodylia. In M. J. Benton (ed.), *The Phylogeny and Classification of the Tetrapods Volume 1: Amphibians, Reptiles, Birds*. Oxford: Clarendon Press, pp. 295–338.

Binford, L. R. 1981. *Bones: Ancient Men and Modern Myths*. New York: Academic Press, p. 320.

Blum, M. and M. Pecha. 2014. Mid-Cretaceous to Paleocene North American drainage reorganization from detrital zircons. *Geology* 42: 607–10.

Bremer, K. 1988. The limits of amino-acid sequence data in angiosperm phylogenetic reconstruction. *Evolution* 42: 795–803.

Bremer, K. 1994. Branch support and tree stability. *Cladistics* 10: 295–304.

Brochu, C. A. 1996. Closure of neurocentral sutures during crocodilian ontogeny: implications for maturity assessment in fossil archosaurs. *Journal of Vertebrate Paleontology* 16: 49–62.

Brochu, C. A. 2001. Crocodylian snouts in space and time: Phylogenetic approaches toward adaptive radiation. *American Zoologist* 41: 564–85.

Brochu, C. A., J. R. Wagner, S. Jouve, C. D. Sumrall, and L. D. Densmore. 2009. A correction corrected: Consensus over the meaning of Crocodylia and why it matters. *Systematic Biology* 58: 537–43.

Brownstein, C. D. (2018). A Tyrannosauroid from the Lower Cenomanian of New Jersey and its evolutionary and biogeographic implications. *Bulletin of the Peabody Museum of Natural History* 59(1): 95–105.

Buffetaut, E. 1983. Wounds on the jaw of an Eocene mesosuchian crocodilian as possible evidence for the antiquity of crocodilian intraspecific fighting behavior. *Paleontologische Zeitschrift* 57: 143–5.

Busby, A. B. 1995. The structural consequences of skull flattening in crocodilians; In J. J. Thomason (ed.), *Functional Morphology in Vertebrate Paleontology*. New York: Cambridge University Press, pp. 173–92.

Cumbaa, S. L., K. Shimada, and T. D. Cook. 2010. Mid-Cenomanian vertebrate faunas of the Western Interior Seaway of North America and their evolutionary, paleobiogeographical, and paleoecological implications. *Palaeogeography, Palaeoclimatology, Palaeoecology* 295: 199–214.

D'Amore, D. C. and R. J. Blumenschine. 2009. Komodo monitor (*Varanus komodoensis*) feeding behavior and dental function reflected through tooth marks on bone surfaces, and the application to ziphodont paleobiology. *Paleobiology* 35(4): 525–52.

D'Amore, D. C., M. Harmon, S. K. Drumheller, and J. J. Testin. 2019. Quantitative heterodonty in Crocodylia: Assessing size and shape across modern and extinct taxa. *PeerJ* 7: e6485.

Dodge, C. F. 1952. Stratigraphy of the Woodbine Formation in the Arlington area. Tarrant County, Texas. *Field and Laboratory* 20(2): 66–78.

Dodge, C. F. 1969. Stratigraphic nomenclature of the Woodbine Formation Tarrant County, Texas. *Texas Journal of Science* 21: 43–62.

Drumheller, S. K. and C. A. Brochu. 2014. A diagnosis of *Alligator mississippiensis* bite marks with comparisons to existing crocodylian datasets. *Ichnos* 21: 131–46.

Drumheller, S. K. and C. A. Brochu. 2016. Phylogenetic taphonomy: a statistical and phylogenetic approach for exploring taphonomic patterns in the fossil record using crocodylians. *Palaios* 31(10): 463–78.

Drumheller, S. K. and E. W. Wilberg. 2020. A synthetic approach for assessing the interplay of form and function in the crocodyliform snout. *Zoological Journal of the Linnean Society* 188(2): 507–21.

Drumheller, S. K., E. W. Wilberg, and R. W. Sadleir. 2016. The utility of captive animals in actualistic research: A geometric morphometric exploration of the tooth row of *Alligator mississippiensis* suggesting ecophenotypic influences and functional constraints. *Journal of Morphology* 277(7): 866–78.

Drumheller, S. K., D'Amore D. C., and Njau, J. K., in press. Taphonomic approaches to bite mark analyses in the fossil record and applications to crocodyliform and broader archosaurian paleobiology. In H. N. Woodward and J. O. Farlow (eds.), *Crocodylian Biology and Archosaur Paleobiology: Studies in Honor of Ruth N. Elsey*. Bloomington, IN: Indiana University Press.

Emerson, B. L., J. H. Emerson, R. E. Akers, and T. J. Akers. 1994. Texas Cretaceous Ammonites and Nautiloids. Houston, TX: Houston Gem & Mineral Society, 439 pp.

Gignac, P. and H. O'Brien. 2016. Suchian feeding success at the interface of ontogeny and macroevolution. *Integrative and Comparative Biology* 56(3): 449–58.

Gignac, P. M. and S. E. Santana. 2016. A bigger picture: Organismal function at the nexus of development, ecology, and evolution: An introduction to the symposium. *Integrative and Comparative Biology* 56(3): 369–72.

Gignac, P. M., S. E. Santana, and H. D. O'Brien. 2019. Ontogenetic inertia explains Neosuchian giants: A case study of *Sarcosuchus imperator* (Archosauria: Suchia). *Journal of Morphology* 280: S65–66.

Goloboff, P. and S. Catalano. 2016. TNT version 1.5, including a full implementation of phylogenetic morphometrics. *Cladistics.* https://doi.org/10.1111/cla.12160

Goloboff, P. A., J. S. Farris, M. Källersjö, et al. 2003. Improvements to resampling measures of group support. *Cladistics* 19: 324–32.

Goloboff, P. A., J. S. Farris, and K. C. Nixon. 2008. TNT, a free program for phylogenetic analysis. *Cladistics* 24: 774–86.

Gradstein, F. M., J. G. Ogg, and A. G. Smith (eds.). 2004. *A Geologic Time Scale.* Cambridge: Cambridge University Press, 500 pp.

Hay, O. P. 1930. *Second Bibliography and Catalogue of the Fossil Vertebrata of North America.* Vol. 1. Washington, DC: Carnegie Institution of Washington Publication, 390, 990 pp.

Head, J. 1998. A new species of basal hadrosaurid (Dinosauria, Ornithischia) from the Cenomanian of Texas. *Journal of Vertebrate Paleontology* 18: 718–38.

Hentz, T. F., W. A. Ambrose, and D. C. Smith. 2014. Eaglebine play of the southwestern East Texas basin: Stratigraphic and depositional framework of the Upper Cretaceous (Cenomanian-Turonian) Woodbine and Eagle Ford Groups. *American Association of Petroleum Geologists Bulletin* 83: 2551–80.

Hua, S., E. Buffetaut, C. Legall, and P. Rogron. 2007. Oceanosuchus boecensis n. gen, n. sp., a marine pholidosaurid (Crocodylia, Mesosuchia) from the Lower Cenomanian of Normandy (western France). *Bulletin de la Société géologique de France* 178(6): 503–13.

Iijima, M. 2017. Assessment of trophic ecomorphology in non-alligatoroid crocodylians and its adaptive and taxonomic implications. *Journal of Anatomy* 231(2): 192–211.

Ikejiri, T. 2012. Histology-based morphology of the neurocentral synchondrosis in *Alligator mississippiensis* (Archosauria, Crocodylia). *The Anatomical Record: Advances in Integrative Anatomy and Evolutionary Biology* 295: 18–31.

Iordansky, N. N. 1973. The skull of the Crocodilia. In C. Gans and T. S. Parsons (eds.), *Biology of the Reptilia*, Vol. 4. New York: Academic Press, pp. 201–62.

Jacobs, L. L. and D. A. Winkler. 1998. Mammals, archosaurs, and the Early to Late Cretaceous transition in north-central Texas; In Y. Tomida, L. J. Flynn, and L. L. Jacobs (eds.), *Advances in Vertebrate Paleontology and Geochronology*. Tokyo: National Science Museum, pp. 253–80.

Johnson, R. O. 1974. Lithofacies and Depositional Environments of the Rush Creek Member of the Woodbine Formation (Gulfian) of North Central Texas. Arlington, TX: University of Texas, 158 pp.

Jouve, S. and N.-E. Jalil. 2020. Paleocene resurrection of a crocodylomorph taxon: Biotic crises, climatic and sea level fluctuations. *Gondwana Research* 85: 1–8.

Jouve, S., D. de Muizon, R. Cespedes-Paz, V. Sossa-Soruco, and S. Knoll. In press. The longirostrine crocodyliforms from Bolivia and their evolution through the Cretaceous-Palaeogene boundary. Zoological Journal of the Linnean Society. https://doi.org/10.1093/zoolinnean/zlaa081

Katsura, Y. 2004, Paleopathology of *Toyotamaphimeis machikanensis* (Diapsida, Crocodylia) from the middle Pleistocene of central Japan. *Historical Biology* 16: 93–7.

Kauffman, E. G. and W. G. E. Caldwell. 1993. The Western Interior Basin in space and time. In E. G. Kauffman and W. G. E. Caldwell (eds.), Evolution of the Western Interior Basin. Geological Society of Canada, Special Papers, 39, pp. 1–30.

Kennedy, W. J. and W. A. Cobban. 1990. Cenomanian ammonite faunas from the Woodbine Formation and lower part of the Eagle Ford Group, Texas. *Palaeontology* 33: 75–154.

Krumenacker, L. J., D. J. Simon, G. Scofield, and D. J. Varricchio. 2016. Theropod dinosaurs from the Albian–Cenomanian Wayan Formation of eastern Idaho. *Historical Biology*. https://doi.org/10.1080/08912963 .2015.1137913

Kuzmin, I. T., P. P. Skutschas, E. A. Boitsova, and H-D. Sues. 2019. Revision of the large crocodyliform *Kansajsuchus* (Neosuchia) from the Late Cretaceous of Central Asia. *Zoological Journal of the Linnean Society* 185: 335–87. https://doi.org/10.1093/zoolinnean/zly027

Langston, W. 1973. The crocodilian skull in historical perspective. In C. Gans and T. Parsons (eds.), *Biology of the Reptilia*, Vol. 4. London: Academic Press, pp. 263–84.

Lee, Y.-N. 1997a. Archosaurs from the Woodbine Formation (Cenomanian) in Texas. *Journal of Paleontology* 71: 1147–56.

Lee, Y.-N. 1997b. Bird and dinosaur footprints in the Woodbine Formation (Cenomanian), Texas. *Cretaceous Research* 18: 849–64.

Lehman, T. M., S. L. Wick, A. A. Brink, and T. A. Shiller, II. 2019. Stratigraphy and vertebrate fauna of the lower shale member of the Aguja Formation (lower Campanian) in West Texas. *Cretaceous Research* 99: 291–314. https://doi.org/10.1016/j.cretres.2019.02.028

Mackness, B. S., J. E. Cooper, C. Wilkinson, and D. Wilkinson. 2010. Paleopathology of a crocodile femur from the Pliocene of eastern Australia. *Alcheringa, An Australasian Journal of Paleontology* 34: 515–21.

Main, D. J. 2013. Appalachian delta plain paleoecology of the Cretaceous Woodbine Formation at the Arlington Archosaur Site, North Texas. Ph.D. dissertation, University of Texas at Arlington, Arlington, Texas, 548 pp.

Main, D. J., C. R. Noto, and D. B. Weishampel. 2014. Postcranial anatomy of a basal hadrosauroid (Dinosauria: Ornithopoda) from the Cretaceous (Cenomanian) Woodbine Formation of North Central Texas. In D. Eberth and D. Evans (eds.), *Hadrosaurs*. Bloomington, IN: Indiana University Press, pp. 77–95.

Martin, J. E., K. Lauprasert, E. Buffetaut, R. Liard, and V. Suteethorn. 2014. A large pholidosaurid in the Phu Kradung Formation of north-eastern Thailand. *Palaeontology* 57: 757–69.

McHenry, C. R., P. D. Clausen, W. J. T. Daniel, M. B. Meers, and A. Pendharkar. 2006. Biomechanics in the rostrum in crocodilians: a comparative analysis using finite-element modeling. *Anatomical Record Part A* 288: 827–49.

McIlhenny, E. A. 1935. *The Alligator's Life History*. Boston, MA: Christopher Publishing House.

McNulty, C. L. and B. H. Slaughter. 1968. Fishbed conglomerate fauna, Arlington Member, Woodbine Formation (Cenomanian) of Texas; In C. F. Dodge (ed.), *Stratigraphy of the Woodbine Formation: Tarrant County, Texas. Fieldtrip Guidebook*. Denver, CO: Geological Society of America, South-Central Section, pp. 68–73.

Meunier, L. M. V. and H. C. E. Larsson. 2017. Revision and phylogenetic affinities of *Elosuchus* (Crocodyliformes). *Zoological Journal of the Linnean Society* 179(1): 169–200.

Meyer, H. 1841. *Pholidosaurus schaumburgensis* ein Saurus aus dem Sandstein der Wald-Formation Nord-Deutschlands. *Neües Jahrbuch für Mineralogie, Geologie und Palaontologie* 4: 443–5.

Njau, J. K. and R. J. Blumenschine. 2006. A diagnosis of crocodile feeding traces on larger mammal bone, with fossil examples from the Plio-Pleistocene Olduvai Basin, Tanzania. *Journal of Human Evolution* 50: 142–62.

Noto, C. R. 2015. Archosaur Localities in the Woodbine Formation (Cenomanian) of North-Central Texas. In Early- and Mid-Cretaceous Archosaur Localities of North-Central Texas. Fieldtrip Guidebook for the 75th Annual Meeting of the Society of Vertebrate Paleontology, Dallas, Texas, C. Noto (ed.), p. 38–51.

Noto, C. R. 2016. New theropods from the Woodbine Formation of Texas: Insights into Cenomanian Appalachian ecosystems. *Journal of Vertebrate Paleontology*, Program and Abstracts: 197.

Noto, C. R., D. J. Main, and S. K. Drumheller. 2012. Feeding traces and paleobiology of a Cretaceous (Cenomanian) Crocodyliform: Example from the Woodbine Formation of Texas. *Palaios* 27: 105–15.

Noto, C. R., T. L. Adams, S. K. Drumheller, and A. H. Turner. 2019. A small enigmatic neosuchian crocodyliform from the Woodbine Formation of Texas. *The Anatomical Record*. https://doi.org/10.1002/ar.24174

Oliver, W. B. 1971. Depositional systems in the Woodbine Formation (Upper Cretaceous), northeast Texas. The University of Texas at Austin, *Bureau of Economic Geology, Report of Investigations* 73, 28 pp.

Owen, R. 1841. On British fossil reptiles: Report of the British Association for the Advancement of Science 11: 60–204.

Pierce, S. E., K. D. Angielczyk, and E. J. Rayfield. 2008. Patterns of morpho-space occupation and mechanical performance in extant crocodilian skulls: A combined geometric morphometric and finite element modeling approach. *Journal of Morphology* 269: 840–64.

Pol, D. and Z. Gasparini. 2009. Skull anatomy of *Dakosaurus andiniensis* (Thalattosuchia Crocodylomorpha) and the phylogenetic position of Thalattosuchia. *Journal of Systematic Palaeontology* 7: 163–97.

Powell, J. D. 1968. Woodbine-Eagle Ford transition, Tarrant Member. In F. Dodge (ed.), *Stratigraphy of the Woodbine Formation: Tarrant County, Texas. Field Trip Guidebook*. Denver, CO: Geological Society of America, South Central Section, pp. 27–43.

Prieto-Márquez, A., G. M. Erickson, and J. A. Ebersole. 2016. Anatomy and osteohistology of the basal hadrosaurid dinosaur *Eotrachodon* from the uppermost Santonian (Cretaceous) of southern Appalachia. *PeerJ* 4: e1872.

Radloff, F. G. T., K. A. Hobson, and A. J. Leslie. 2012. Characterizing ontogen-etic niche shifts in Nile crocodile using stable isotopes ($\delta13$ C, $\delta15$ N) analyses of scite keratin. *Isotopes in Environmental and Health Studies* 48(3): 43956.

Ristevski, J., M. T. Young, M. B. de Andrade, A. K. Hastings. 2018. A new species of *Anteopthalmosuchus* (Crocodylomorpha, Goniopholididae) from the Lower Cretaceous of the Isle of Wight, United Kingdom, and a review of the genus. *Cretaceous Research* 84: 340–83.

Ross C. A. and Magnusson W. E. 1989. Living crocodilians. In C. A. Ross (ed.), *Crocodiles and Alligators*. New York: Facts on File, pp. 58–73.

Russell, D. A. 1988. A checklist of North American marine Cretaceous vertebrates including fresh water fishes. *Occasional Paper of Tyrrell Museum of Palaeontology* 4: 1–57.

Sadleir, R. W., and P.J. Makovicky. 2008. Cranial shape and correlated characters in crocodilian evolution. Journal of Evolutionary Biology 21:1578–1596.

Salas-Gismondi R., J. J. Flynn, P. Baby, et al. 2015. A Miocene hyperdiverse crocodylian community reveals peculiar trophic dynamics in proto-Amazonian mega-wetlands. *Proceedings of the Royal Society B: Biological Sciences* 282: 20142490.

Schmidt, K. P. 1944. Crocodiles. *Fauna* 6: 67–72.

Schwartz D., M. Raddatz, and O. Wings. 2017. *Knoetschkesuchus langenbergensis* gen. nov. sp. nov., a new atoposaurid crocodyliform from the Upper Jurassic Langenberg Quarry (Lower Saxony, northwestern Germany), and its relationships to *Theriosuchus*. *PLoS One* 12: e0160617.

Sereno, P. C. and H. C. E. Larsson. 2009. Cretaceous crocodyliforms from the Sahara. *ZooKeys* 28: 1–143.

Sereno P. C., H. C. E Larsson, C. A. Sidor, and B. Gado. 2001. The giant crocodyliform *Sarcosuchus* from the Cretaceous of Africa. *Science* 294: 1516–19.

Strganac, C. 2015. *Field Trip Overview. In Early- and Mid-Cretaceous Archosaur Localities of North-Central Texas.* Fieldtrip Guidebook for the 75th Annual Meeting of the Society of Vertebrate Paleontology, Dallas, Texas, C. Noto (ed.), p. 2–4.

Turner, A. H. 2015. A review of *Shamosuchus* and *Paralligator* (Crocodyliformes, Neosuchia) from the Cretaceous of Asia. *PLoS ONE* 10: e0118116.

Ullmann, P. V., D. J. Varricchio, and M. J. Knell. 2012. Taphonomy and taxonomy of a vertebrate microsite in the mid-Cretaceous (Albian–Cenomanian) Blackleaf Formation, southwest Montana. *Historical Biology* 24: 311–28.

Vasconcellos, F. M. and I. S. Carvalho. 2010. Paleoichnological assemblage associated with *Baurusuchus salgadoensis* remains, a Baurusuchidae Mesoeucrocodylia from the Bauru Basin, Brazil (Late Cretaceous). *Bulletin of the New Mexico Museum of Natural History and Science* 51: 227–37.

Webb, G. J. W., S. C. Manolis, and R. Buckworth. 1982. *Crocodylus johnstoni* in the McKinlay River Area, NTI Variation in the diet, and a new method of assessing the relative importance of prey. *Australian Journal of Zoology* 30(6): 877–99.

Weishampel, D. B., P. M. Barrett, R. A. Coria, et al. 2004. Dinosaur distribution. In D. B. Weishampel, P. Dodson, and H. Osmólska (eds.), *The Dinosauria*, 2nd ed. Berkeley, CA: University of California Press, pp. 517–606.

Whetstone, K. and P. Whybrow. 1983. A 'cursorial' crocodilian from the Triassic of Lesotho (Basutoland), southern Africa. *Occasional Publications of the Museum of Natural History of the University of Kansas* 106: 1–37.

Wilberg, E. W. 2017. Investigating patterns of crocodyliform cranial disparity through the Mesozoic and Cenozoic. *Zoological Journal of the Linnean Society* 181(1): 189–208.

Williamson, W. E. 1996.? *Brachychampsa sealeyi*, sp. nov., (Crocodylia, Alligatoroidea) from the Upper Cretaceous (lower Campanian) Menefee Formation, northwestern New Mexico. *Journal of Vertebrate Paleontology* 16: 421–31.

Wu, X.-C., A. P. Russell, and S. L. Cumbaa. 2001. *Terminonaris* (Archosauria: Crocodyliformes): New material from Saskatchewan, Canada, and comments on its phylogenetic relationships. *Journal of Vertebrate Paleontology* 21: 492–514.

Young, M. T., A. K. Hastings, R. Allain, and T. J. Smith. 2016. Revision of the enigmatic crocodyliform Elosuchus felixi de Lapparent de Broin, 2002 from the Lower Upper Cretaceous boundary of Niger: Potential evidence for an early origin of the clade Dyrosauridae. *Zoological Journal of the Linnaean Society*. https://doi.org/10.1111/zoj.12452

Zanno, L. E. and P. J. Makovicky. 2013. Neovenatorid theropods are apex predators in the Late Cretaceous of North America. *Nature Communications* 4: 2827.

Acknowledgments

We thank the Huffines family and R. Kimball of Johnson Development, who provided access to the property on which the AAS is situated. A. Sahlstein, P. Kirchoff, and B. Walker first discovered the site, and a dedicated team of volunteers helped excavate and prepare much of the material over the past 10 years. Thanks to the Perot Museum of Nature and Science, Southern Methodist University, and the Witte Museum for access to research collections. We thank S. Jouve and an anonymous reviewer for their valuable comments that helped improve this manuscript. C. Sumrall prepared additional specimens and provided access to photographic equipment during the COVID-19 quarantine. Taxon silhouettes come from Phylopic (phylopic.org), and were created by S. Werning (Wonder Women) and Smokeybjb (crocodyliform). These images are used under a Creative Commons Attribution-Non-Commercial-Share Alike 3.0 license. This study was supported by funding from the National Geographic Society Conservation Trust grant #C325-16 to C. Noto, National Science Foundation DUE-IUSE GP-IMPACT grant #1600376 to L. McKay and S. Horn, as well as our contributors on Experiment.com.

Cambridge Elements ≡

Elements of Paleontology

Editor-in-Chief
Colin D. Sumrall
University of Tennessee

About the Series
The Elements of Paleontology series is a publishing collaboration between the Paleontological Society and Cambridge University Press. The series covers the full spectrum of topics in paleontology and paleobiology, and related topics in the Earth and life sciences of interest to students and researchers of paleontology.

The Paleontological Society is an international nonprofit organization devoted exclusively to the science of paleontology: invertebrate and vertebrate paleontology, micropaleontology, and paleobotany. The Society's mission is to advance the study of the fossil record through scientific research, education, and advocacy. Its vision is to be a leading global advocate for understanding life's history and evolution. The Society has several membership categories, including regular, amateur/avocational, student, and retired. Members, representing some 40 countries, include professional paleontologists, academicians, science editors, Earth science teachers, museum specialists, undergraduate and graduate students, postdoctoral scholars, and amateur/avocational paleontologists.

Paleontological
S O C I E T Y

Cambridge Elements $\equiv$

Elements of Paleontology

Elements in the Series

Printed in the United States
by Baker & Taylor Publisher Services